Greenhouse Emission Reductions and Natural Gas Vehicles: A Resource Guide on Technology Options and Project Development

September 2002

Prepared for the National Energy Technology Laboratory (NETL)
626 Cochrans Mill Road
P.O. Box 10940
Pittsburgh, PA 15236-0940
www.netl.doe.gov

by

Science Applications International Corporation (SAIC)
Climate Change Services
8301 Greensboro Drive, E-5-7
McLean, Virginia 22102
www.saic.com

With contributions from:
Orestes Anastasia, Nancy Checklick, Vivianne Couts, Julie Doherty,
Jette Findsen, Laura Gehlin, and Josh Radoff
SAIC

Reviewed by:
Richard Bechtold, QSS, Inc.
Jim Ekmann, NETL
George Lee, NETL

Disclaimer

This report was prepared as an account of work sponsored by an agency of the United States Government. Neither the United States Government nor any agency thereof, nor any of their employees, makes any warranty, express or implied, or assumes any legal liability or responsibility for the accuracy, completeness, or usefulness of any information, apparatus, product, or process disclosed, or represents that its use would not infringe upon privately owned rights. Reference herein to any specific commercial product, process, or service by trade name, trademark, manufacturer, or otherwise, does not necessarily constitute or imply its endorsement, recommendation, or favoring by the United States Government or any agency thereof. The views and opinions of authors expressed herein do not necessarily state or reflect those of the United States Government or any agency thereof.

Table of Contents

Acknowledgements

The authors express their appreciation to Jim Ekmann of the National Energy Technology Laboratory for his ideas, inspiration and support. We also thank Rich Bechtold and George Lee who provided insightful review comments and helped refine some of the material presented here. Finally, we thank Marcy Rood of the U.S. Department of Energy's Clean Cities Program for her encouragement and support for this effort.

Acronyms Used in this Report

ACP	Alternative Compliance Plan
AFV	Alternative Fuel Vehicle
AIJ	Activities Implemented Jointly
ANL	Argonne National Laboratory
AT PZEV	Advanced Technology Partial Zero-Emission Vehicle
CAA	Federal Clean Air Act
CAFE	Corporate Average Fuel Economy
CARB	California Air Resources Board
CARFG2	California Phase 2 Reformulated Gasoline
CCX	The Chicago Climate Exchange
CD	Conventional Diesel
CDM	Clean Development Mechanism
CFF	Clean Fuel Fleet
CFVs	Clean Fuel Vehicles
CH_4	Methane
CIDI	Compression Ignition, Direct Injection
CMSA	Consolidated Metropolitan Statistical Area
CO	Carbon Monoxide
CO_2	Carbon Dioxide
CO_2E	Carbon Dioxide Equivalent
COP	Conference of Parties
CNG	Compressed Natural Gas
CSDA	Center for Sustainable Development in the Americas
CV	Conventional Vehicle
DOE	U.S. Department of Energy
DOT	U.S. Department of Transportation
E10	A Mixture of 10% Ethanol and 90% Gasoline
E85	A Mixture of 85% Ethanol and 15% Gasoline
EIA	Energy Information Administration
EPAct	Energy Policy Act of 1992
E.O.	Executive Order
ERUPT	Emission Reduction Unit Procurement Tender, The Netherlands
ETBE	Ethyl Tertiary Butyl Ether
FAA	Federal Aviation Administration
FFV	Fuel Flexible Vehicle
FRFG2	Federal Phase 2 Reformulated Gasoline
GHG	Greenhouse Gas
GI	Grid Independent
GMC	General Motors Corporation
GREET	Greenhouse Gases, Regulated Emissions, and Energy Use in Transportation
GSA	General Services Administration
GTL	Gas-to-Liquid
GV	Gasoline Vehicle
GWP	Global Warming Potential
HC	Hydrocarbon
HFCs	Hydroflourocarbons
HOV	High-Occupancy Vehicle
ICE	Internal Combustion Engine
ICLEI	International Council for Local Environmental Initiatives
ILEAV	Inherently Low-Emission Airport Vehicle
IPCC	Intergovernmental Panel on Climate Change

JI	Joint Implementation
LDV	Light-Duty Vehicle
LEV	Low-Emission Vehicle
LNG	Liquid Natural Gas
LPG	Liquid Petroleum Gas
Mpg	Miles per gallon
MSA	Metropolitan Statistical Area
MTBE	Methyl Tertiary Butyl Ether
MY	Model Year
M85	A Mixture of 85% Methanol and 15% Gasoline
NETL	National Energy Technology Laboratory
NG	Natural Gas
NGV	Natural Gas Vehicle
NMOG	Non-Methane Organic Gas
NO_x	Nitrous Oxides (unspecified)
N_2O	Nitrous Oxide
OEM	Original Equipment Manufacturer
OMB	Office of Management and Budget
PC	Passenger Car
PCF	Prototype Carbon Fund
PFCs	Perflourocarbons
PERT	Pilot Emissions Reduction Program, Canada
PRD	Pressure Relief Devices
PZEV	Partial Zero-Emission Vehicle
P&LG	Private and Local Government
SF_6	Sulphur Hexaflouride
SIDI	Spark Ignition, Direct Injection
SIP	State Implementation Plan
SULEV	Super-Ultra-Low-Emission Vehicle
SUV	Sport Utility Vehicle
S&FP	State and Alternative Fuel Provider
tCO_2E	Tons of Carbon Dioxide Equivalent
TSP	Total Suspended particulates
TLEV	Transitional Low Emissions Vehicle
ULEV	Ultra-Low-Emission Vehicle
USIJI	U.S. Initiative on Joint Implementation
UNCED	United Nations Conference on Environment and Development
UNFCCC	United Nations Framework Convention on Climate Change
WBCSD	World Business Council for Sustainable Development
WRI	World Resources Institute
ZEV	Zero-Emission Vehicle

Overview

The transportation sector accounts for a large and growing share of global greenhouse gas (GHG) emissions. Worldwide, motor vehicles emit well over 900 million metric tons of carbon dioxide (CO_2) each year, accounting for more than 15 percent of global fossil fuel-derived CO_2 emissions.[1] In the industrialized world alone, 20-25 percent of GHG emissions come from the transportation sector. The share of transport-related emissions is growing rapidly due to the continued increase in transportation activity.[2] In 1950, there were only 70 million cars, trucks, and buses on the world's roads. By 1994, there were about nine times that number, or 630 million vehicles. Since the early 1970s, the global fleet has been growing at a rate of 16 million vehicles per year. This expansion has been accompanied by a similar growth in fuel consumption.[3] If this kind of linear growth continues, by the year 2025 there will be well over one billion vehicles on the world's roads.[4]

In a response to the significant growth in transportation-related GHG emissions, governments and policy makers worldwide are considering methods of addressing this trend. However, due to the particular make-up of the transportation sector, regulating and reducing emissions from this sector poses a significant challenge. Unlike stationary fuel combustion, transportation-related emissions come from dispersed sources. Only a few point-source emitters, such as oil/natural gas wells, refineries, or compressor stations, contribute to emissions related to the transportation sector. The majority of transport-related emissions come from the millions of vehicles traveling the world's roads. As a result, successful GHG mitigation policies must find ways to target all of these small, non-point source emitters, either through regulatory means or through various incentive programs. To increase their effectiveness, policies to control emissions from the transportation sector often utilize indirect means to reduce emissions, such as requiring specific technology improvements or an increase in fuel efficiency. Site-specific project activities can also be undertaken to help decrease GHG emissions, although the use of such measures is less common. These activities include switching to less GHG-intensive vehicle options, such as natural gas vehicles (NGVs). As emissions from transportation activities continue to rise, it will be necessary to promote both types of abatement activities in order to reverse the current emissions path. This Resource Guide focuses on site- and project-specific transportation activities.

Over the last decade, efforts to reduce GHG emissions in the U.S. have led to the creation of a number of voluntary programs for registering and crediting project-specific

---

[1] World Resources Institute, "Proceed With Caution: Growth in the Global Motor Vehicle Fleet," http://www.wri.org/trends/autos.html.

[2] "Good Practice Greenhouse Abatement Policies: Transport Sector," OECD and IEA Information Papers prepared for the Annex I Expert Group on the UNFCCC, OECD and IEA (Paris, November 2000). Emissions exclude land-use change and forestry, and bunker fuels. Annex I countries are those countries that have undertaken binding emission reduction targets under the Kyoto Protocol of the United Nations Framework Convention on Climate Change (UNFCCC).

[3] American Automobile Manufacturers Association (AAMA), "World Motor Vehicle Data 1993," AAMA (Washington, D.C., 1993), p. 23, and American Automobile Manufacturers Association (AAMA), "Motor Vehicle Facts and Figures 1996," AAMA (Washington, D.C., 1996).

[4] World Resources Institute, "Proceed With Caution: Growth in the Global Motor Vehicle Fleet," http://www.wri.org/trends/autos.html.

GHG reduction activities undertaken by individual project developers. Similarly, several international programs have been implemented, including efforts that allow for trading in GHG emission reduction activities. As a result, a small but growing market for the trade in GHG emission reduction credits has emerged, creating an additional incentive for project developers in the transportation sector to undertake GHG reduction projects. Given that certain applications of NGVs emit less GHG emissions compared to conventional vehicles, projects that lead to the introduction of NGVs can register with the many voluntary GHG reporting programs and could potentially be able to sell the associated GHG reduction credits on the market. However, to participate in these efforts, project developers must be familiar with the procedures for developing and estimating the GHG emissions benefits resulting from the various types of projects.

To date, only a few projects deploying NGV technologies have been developed and implemented with the explicit intent of reducing GHG emissions and participating in international GHG reduction initiatives. Therefore, experience with quantifying, evaluating, and verifying GHG emission reductions from natural gas vehicle projects is almost non-existent. This is a problem as there are many issues unique to the transportation sector, which should be resolved before adequate guidelines can be developed for evaluating transportation-related projects. Issues that will require further analysis and guidance include:

1. Methods for accurately **estimating emission reductions** for a dispersed number of sources;

2. Procedures for **determining project boundaries** and relevant **GHG emission sources**;

3. Options for **minimizing transaction costs** of validating, monitoring, verifying, and certifying potential emission reductions; and

4. Guidance on using a **full fuel-cycle or tailpipe emission analysis** to estimate project emissions.

The main purpose of this manual is to provide information on quantifying and documenting GHG emission reductions from NGV projects. Moreover, to provide potential project developers with an overview of project opportunities, the manual also includes information on NGV technology cost and availability and discusses the future of the alternative fuel vehicle (AFV) industry as a whole.

Chapter 1 of this report provides an outline of NGV technology availability, including information on safety, cost, vehicle types, and refueling infrastructure. The purpose of this chapter is to provide an understanding of the availability of NGV technologies in the short-term and describe worldwide deployment.

Chapter 2 describes domestic and international regulatory frameworks for NGVs. It provides information on existing and pending regulatory activities under which an NGV project developer could receive credits for initiating an NGV GHG emission reduction project. Domestic laws and regulations include the Federal Energy Policy Act of 1992, the Federal Clean Air Act, as well as a variety of state policy initiatives that favor alternative fuel vehicles and NGVs. The chapter also addresses a variety of domestic and international programs promoting the development of GHG reduction projects and nascent emissions trading schemes.

Chapter 3 examines the GHG emissions associated with NGV use and reviews recent literature and models for estimating NGV-associated GHG emissions. The chapter also summarizes current projects deploying NGVs and discusses the differences between quantifying tailpipe vs. full fuel cycle emissions from NGV use. Chapter 3 concludes with a discussion of the barriers to the implementation of transportation-related GHG mitigation projects and provides suggestions for measures to overcome such barriers.

A case study describing the steps necessary to quantify and document GHG emission reductions from an NGV project applying criteria under the United Nations Framework Convention on Climate Change (UNFCCC) is presented in Chapter 4. This chapter provides a step-by-step description of developing emission baselines and estimating net GHG emission benefits of a hypothetical project that replaces gasoline-fueled taxis with compressed natural gas-fueled taxis. The case study includes an analysis of both tailpipe and full fuel cycle emissions of the project.

1 Natural Gas Vehicle Technology Options

1.1 Introduction

The objective of this chapter is to familiarize the reader with the different NGV technology options currently available and provide an overview of the status of NGV deployment worldwide.[5] Detailed background information, such as safety, cost, and infrastructure availability of NGVs is presented to enable project developers and other entities to evaluate the GHG reduction benefits of a potential GHG reduction project. The procedures for baseline development—a necessary prerequisite for measuring emissions reduction benefits—is discussed in more detail in Chapters 3 and 4.

Natural Gas as a Transportation Fuel

Natural gas is a mixture of gaseous hydrocarbons, composed primarily of methane (CH_4), but also of smaller amounts of ethane, propane, butane, carbon dioxide, and other trace gases; the specific mixture varies by region. Natural gas is produced primarily from gas wells (disassociated) or in conjunction with crude oil production (associated), but can also be produced as a byproduct of landfill and coal mining operations.

At atmospheric pressure, the volumetric energy density of natural gas (the amount of energy contained per unit volume) is too low to warrant use in the relatively small fuel tanks of motor vehicles. Thus, in natural gas vehicles, the gas is either compressed (put under pressure—usually to between 2000 and 3600 pounds per square inch (psi)), or liquefied by reducing its temperature to negative 260°F at atmospheric pressure, to increase the amount of energy that can be stored in a fuel tank. As much as any of the design elements of the engine itself, the storage and safety issues associated with compressing and liquefying the gas have presented some of the greatest challenges in the development and marketing of the technology. However, these challenges—described below—have been duly addressed and enough operating experienced has been gained to place NGVs in the realm of the commercially viable. The basic characteristics of natural gas, which differ from liquid gasoline and therefore present a unique set of challenges with regard to vehicle design, are listed in Table 1-1.

---

[5] This chapter focuses on NGV technologies and deployment in the U.S. market. NGVs are also used in other countries, such as Argentina, Brazil, Canada, Egypt, Italy and Russia. Please see Section 1.5 of this chapter for more information on vehicle use in these countries.

Table 1-1.   Characteristics of Natural Gas for Transportation Fuels

Characteristic	Compressed Natural Gas (CNG)	Liquefied Natural Gas (LNG)
Chemical Structure	CH$_4$	CH$_4$
Primary Components	Methane compressed to 2000 to 3600 PSI	Methane that is cooled to –260 degrees F
Main Fuel Source	Underground reserves	Underground reserves
Energy Content per Gallon	29,000 Btu	73,500 Btu
Energy Ratio Compared to Gasoline by Volume	1 to 3.94 or 25% at 3000 psi	1 to 1.55 or 66%
Liquid or Gas	Gas	Liquid

Source: Alternative Fuels Data Center (AFDC), http://www.afdc.doe.gov/questions.html.

In the U.S., compressed natural gas (CNG) is the most commonly used form of natural gas for NGV vehicles although heavy-duty vehicle fuel markets are developing rapidly for liquefied natural gas (LNG). The reason for CNG's early dominance is due to the fact that in order to maintain LNG as a liquid, it must remain below -117°F, which presents some difficult technical challenges. In terms of operation, LNG differs from CNG in the following ways:

- LNG is stored at low pressure (typically less than one-tenth of that of CNG);

- The liquid composition of LNG allows for economical transportation via trucks, railcars, barges, or ships; and

- The majority of the higher hydrocarbons and virtually all of the contaminants present in CNG are removed when LNG is made.

NGV Engine and Fuel System Description

There is very little difference between a gasoline and natural gas vehicle. NGVs use the same basic engine, with minor changes in compression ratio, ignition timing, and the emission control system. The fuel system normally consists of one or more high pressure CNG tanks (stored either on the roof, on the vehicle undercarriage or in the rear of the vehicle -- pressurized up to 3600 psi), regulators to reduce the pressure for use in the engine, and injectors or a mixing device to meter the natural gas into the engine.[6] As a result of the similarity between NGVs and conventional vehicles, performance tends to be very similar while emissions are lower. Conversion of a gasoline vehicle to a CNG vehicle requires the installation of fuel storage tank(s), a refueling connection, a pressure regulator, and a fuel metering system (most typically a computer-controlled mixing device installed in the intake system).

---

[6] Correspondence with Hank Seiff, Director of Technology, Natural Gas Vehicle Coalition, August, 2002.

NGV Operational Issues

There are a number of operational issues for NGVs that differ from conventional gasoline- or diesel-powered vehicles. The first is vehicle range. Due to the lower energy density per unit volume (BTU/gallon) of natural gas (both CNG and LNG) compared to gasoline and diesel, NGVs must possess extra fuel storage capacity in order to achieve the same range as conventional vehicles. This presents a trade-off in vehicle design between range and available cargo space.

Second, due to the cleaner burning properties of natural gas, vehicle maintenance tends to be reduced (fewer oil changes needed) offset somewhat by reduced brake pad life (typically only for heavy-duty vehicles) and periodic tank inspections (CNG vehicles).

Third, achieving the same refueling rates as gasoline and diesel requires an added amount of cost and energy. Thus, there are typically two refueling options available for NGVs: slow-fill and fast-fill. Slow-fill systems are simpler in design and cost less than fast-fill stations. However, slow-fill stations typically require six to twelve hours to refuel vehicles compared to the two to five minutes needed with fast-fill systems. Slow-fill stations attach the vehicle directly to a compressor and have little or no storage capacity. These stations are used primarily for fleet vehicles that can remain idle in a single location over a longer period of time. Publicly available NGV stations are almost always fast-fill stations with high-pressure storage for faster refueling. Fast-fill CNG refueling stations perform gas compression, drying, and filtration, storage, and dispensing. The gas compressors are expensive and consume significant electric or gas engine power. There are several fast-fill CNG station designs that can include smaller compressors and larger gas storage tanks or larger compressors with reduced storage capacity. The selection generally is driven by the vehicle refueling schedule requirements. CNG refueling dispensers are similar to gasoline or diesel dispensers, except that the nozzles have positive-connect pressure fittings.[7]

Natural Gas Vehicle Types

Natural gas vehicles are dedicated, bi-fuel, or dual-fuel vehicles and are available as light-, medium-, and heavy-duty vehicles. Dedicated vehicles run only on natural gas (most heavy-duty vehicles and some light- and medium-duty vehicles). Bi-fuel vehicles run on either natural gas or gasoline (light- and medium-duty vehicles). Dual-fuel vehicles operate on both natural gas and diesel fuel (heavy-duty vehicles). The advantage of bi-fuel and dual-fuel vehicles is that the operating range is extended, and bi-fuel vehicles can continue to be driven if no natural gas refueling station is available. However, the local pollutant and GHG emission benefits of a dedicated NGV are greater than bi-fuel or dual-fuel NGVs.[8]

---

[7] California Energy Commission, "ABCs of AFVs, A Guide to Alternative Fuel Vehicles," Fifth Edition, November 1999.

[8] With respect to local air pollutants (non-methane hydrocarbons, carbon monoxide, oxides of nitrogen, and particulate matter), exhaust emissions from NGVs are inherently lower and easier to control than gasoline-powered vehicles. This is illustrated by the fact that NGVs have attained California SULEV emissions for many years before any gasoline vehicles did. Dedicated natural gas heavy-duty vehicles typically have half the oxides of nitrogen emissions and less than five percent of the particulate matter emissions of comparable diesel vehicles. In addition, NGVs also emit significantly lower amounts of greenhouse gases and toxins. Dedicated NGVs produce little or no evaporative emissions during fueling and use, while evaporative and fueling emissions account for at least 50 percent of a gasoline vehicle's total hydrocarbon emissions.

1.2   Light-Duty Vehicles

Light-duty vehicles are classified as having a gross vehicle weight of less than 8,500 pounds (3,850 kg).   Typically, passenger cars, small vans, and small trucks are considered to be light-duty vehicles. A typical light-duty CNG vehicle has a driving range of 120 to 180 miles. Light-duty NGVs are typically best suited for fleet use, because this application allows them to be returned to a central location for refueling. Fleet vehicles currently make up the majority of light-duty NGVs on the road today in the U.S., as well as abroad in Argentina and Canada.

As Table 1-2 indicates, all the major U.S. automobile manufacturers (Ford, General Motors, and DaimlerChrysler) offer light-duty NGVs for sale. For model years 2000-2001, there are five manufacturers offering NGVs to U.S. consumers (the three U.S. automakers plus the U.S. operations of Honda and Toyota). All of the NGVs offered by these manufacturers are fueled by CNG.

Table 1-2. Light-Duty NGV Manufacturers

Manufacturer	Body Type	Vehicle Type	Fuel Type
American Honda Motor Co. Inc.	Sedan	Dedicated	CNG
DaimlerChrysler	Van, Wagon	Dedicated	CNG
Ford Motor Co.	Sedan, Pickup, Van, Wagon	Dedicated, Dual-fuel	CNG
General Motors Corp.	Sedan, Pickup	Dual-fuel	CNG
Toyota Motor Sales, U.S.A., Inc.	Sedan	Dedicated	CNG

Source: *1999-2000 Natural Gas Vehicle Purchasing Guide*, Natural Gas Vehicle Coalition (NGVC), http://www.ngvc.org.

1.3   Medium-Duty and Heavy-Duty Vehicles

Medium-duty vehicles are classified as having a gross vehicle weight of between 8,500 pounds and 14,000 pounds (3,850 and 6,350 kg). Medium-duty vehicles typically include trucks, vans, cargo vehicles, shuttle buses, and street sweepers. Heavy-duty vehicles are classified as having a gross vehicle weight of greater than 14,000 pounds (6,350 kg), and include large trucks, transit buses, and school buses. Trucks are suitable for both CNG and LNG use because they have high fuel consumption rates, which reduce the payback time. Some of the country's delivery fleets currently using NGVs include United Parcel Service and the U.S. Postal Service.

Two types of engine operating cycles are currently being used for heavy-duty CNG engines. The first is spark-ignited which uses a spark plug to ignite the natural gas fuel mixture in the combustion chamber, similar to a light-duty automobile engine.   The second is compression pilot ignition. This technology injects a small amount of diesel along with natural gas into the combustion chamber.   The heat generated by compressing this mixture ignites the diesel fuel that in turn ignites the natural gas mixture, operating much like a conventional diesel engine.[9]

Most manufacturers of diesel engines now offer comparable natural gas models. Table 1-3 provides a list of manufacturers offering medium- and heavy-duty NGVs.

---

[9] California Energy Commission, "ABCs of AFVs: A Guide to Alternative Fuel Vehicles," Fifth Edition (November 1999).

Table 1-3.   Medium- and Heavy-Duty NGV Manufacturers

Manufacturer	Vehicle Types	Fuel Types
Blue Bird Corporation	Bus	CNG, LNG
Champion Bus, Inc.	Bus	CNG
ElDorado National	Bus	CNG, LNG
Freightliner	Bus, Truck	CNG, LNG
Mack Trucks, Inc.	Truck, Refuse Hauler	CNG, LNG
Neoplan USA Corp.	Bus	CNG, LNG
New Flyer of America	Bus	CNG
North American Bus Industries	Bus	CNG
Nova Bus	Bus	CNG, LNG
OmniTrans Distributing	Bus, Truck	CNG, LNG
Orion Bus Industries	Bus	CNG
Peterbilt Motors Co.	Refuse Truck	CNG, LNG
Spartan Motors Chassis, Inc.	Bus	CNG, LNG
Thomas Built Buses	Bus	CNG
Western Star Trucks	Truck	CNG, LNG

Source: 1999-2000 Natural Gas Vehicle Purchasing Guide, Natural Gas Vehicle Coalition (NGVC), http://www.ngvc.org.

1.4   Natural Gas Vehicle, Fuel, and Infrastructure Cost

The cost of NGVs will differ depending on whether the vehicle is a dedicated or duel fueled vehicle from an original equipment manufacturer (OEM) or it is converted from a gasoline-or diesel-powered vehicle.   Original equipment NGVs are typically more expensive than their gasoline-or diesel-powered counterparts.   DaimlerChrysler, for example, charges about $4,000 more for a light-duty natural gas vehicle compared with a comparable conventional gasoline model.[10]   General Motors Corporation (GMC) charges approximately $3,700 more than a gasoline vehicle.[11]   With more vehicles coming on the market, this cost-differential is expected to decrease. With regard to converting gasoline-powered vehicles to NGVs, the cost penalty is slightly lower. The price of converting a vehicle to natural gas-use currently ranges between $2,500 and $4,000.[12]   This figure is based on a variety of factors, including vehicle type, number of fuel tanks, and labor and installation costs. The prices of heavy-duty natural gas engines vary. Because of the substantial premium for development costs, prices for heavy-duty natural gas engines are nearly double that of a comparable diesel engine. The incremental cost for each heavy-duty CNG vehicle can range from $20,000 for a fleet of small buses to $60,000 for a large unique CNG demonstration truck.[13]

Although, the cost of NGV technologies is higher than conventional vehicle technologies, the lower natural gas price can offset the economic disadvantage caused by the high equipment cost. Table 1-4 provides a cost comparison of gasoline and CNG. The table illustrates that the price of gasoline is on the order of two to four times as expensive as natural gas, and experiences a greater degree of price instability.

---

[10] California Energy Commission, "NGV–Fuel and Vehicle History and Characteristics," http://www.energy.ca.gov.

[11] Cost figures may vary depending on the vehicle and the engine type.

[12] Id.

[13] California Energy Commission, "ABCs of AFVs: A Guide to Alternative Fuel Vehicles," Fifth Edition (November 1999).

Table 1-4.	Cost Comparison between Gasoline and CNG (1990-2000)					
Year	Gasoline ($/gallon)	Annual Gasoline Price Change (%)	Natural Gas			Annual CNG Price Change (%)
			($/1,000 Cu. Ft.)	($/gallon Gasoline Equivalent)*	x less expensive than gasoline	
1990	1.22	NA	3.39	0.362	3.369	NA
1991	1.20	-1.6	3.96	0.423	2.836	16.8
1992	1.19	-.08	4.05	0.433	2.750	2.3
1993	1.17	-1.7	4.27	0.456	2.565	5.4
1994	1.17	0	4.11	0.439	2.665	-3.7
1995	1.21	3.4	3.98	0.425	2.846	-3.2
1996	1.29	6.6	4.34	0.464	2.782	9
1997	1.29	0	4.44	0.474	2.719	2.3
1998	1.12	13.2	4.59	0.490	2.284	3.4
1999	1.22	8.9	4.34	0.464	2.631	-5.4
August 2000	1.56	27.9	NA	NA	NA	NA

* Assumes that the volumetric energy density of Gasoline is 837.6 times that of Natural Gas at Atmospheric Pressure.

Sources: Energy Information Administration, *Monthly Energy Review*, Table 9.4, Washington, DC, 2000. Energy Information Administration, *Natural Gas Annual, 1999*, Table 95, Washington, DC, 2000.

In addition to considering the costs of vehicle equipment and fuel supplies, project developers should also take into account the cost of installing refueling infrastructure. A 1997 estimate for installing fast-fill compressor facilities for a small private or public fleet of about ten vehicles in California ranged between $180,000 to $250,000. [14]

1.5 NGV Vehicle Maintenance, Infrastructure and Safety

The body, engine, and overall structure of NGVs are similar to conventional fuel vehicles, so that maintenance requirements are essentially are virtually the same for all vehicle components, with the exception of the fuel delivery system. CNG cylinders and LNG tanks should be inspected periodically, and the former should be re-certified to maintain standards. Re-certification is a process of examining the cylinders for manufacturing defects, cracks, or other signs of wear. Gasoline fuel systems on dual-fuel vehicles should be run at least once per week to prevent drying and cracking of the gasoline elastomers of the gasoline system. Due to the clean burning properties of natural gas, engine oil will not appear dirty. However, engine oil breaks down over time and should be changed at manufacturer-recommended intervals. Furthermore, in the case of LNG vehicles, the fuel is cooled cryogenically to -260°F, which can cause cryogenic burning (freezing of the skin) upon contact. This presents the need for additional safety training for handlers, and methane gas detectors to detect leaks.

Re-Fueling Station Requirements

Unlike gasoline or diesel, which must be processed from crude oil in large, complex refineries, natural gas requires very little processing to make it suitable for use as a transportation fuel. After water vapor, sulfur, and heavy hydrocarbons are removed,

---

[14] Ibid.

natural gas can be transported via pipeline directly to a NGV refueling station where it is compressed for use. Alternatively, it can be liquefied and stored at the refueling station or transported in liquid form by truck to the station. The slow-and fast-fill delivery mechanisms are described in Section 1, NGV Operational Systems above.

Basic Refueling Station

The basic NGV refueling station is made up of the following typical components: compressor, controls, ground storage, dispensing, and metering. Gas is transported through pipelines in a non-compressed state. However, prior to refueling, the gas must be compressed to approximately 3,600 psi for use in CNG vehicles. In the case of LNG, the gas is liquefied, and must be kept in a liquid state until the time of refueling. Tanks made of a double-wall construction, are used to store LNG. [15] In the case of slow-filled NGVs, there is no need for on-site storage as the natural gas can be routed directly from the pipeline to the compressor station.

Refueling Safety

The overall safety record of natural gas as a vehicle fuel is good. Although natural gas becomes flammable in air, at concentrations of 5 to 15 percent, it has certain characteristics that facilitate safety. Natural gas is a vapor rather than a liquid. Unlike liquid fuels, which will pool on the ground when leaked or spilled, natural gas will dissipate into the atmosphere because it is lighter than air. The ignition temperature of natural gas is also much higher than gasoline; 1,200 degrees Fahrenheit compared to 600 degrees Fahrenheit. Even so, leak detection is an important component of any natural gas system. Because natural gas is odorless, odorants are added to facilitate leak detection. During the liquefaction process, however, these odorants are removed, making detection of LNG leaks more challenging.

To improve safety from leaks or explosions, natural gas storage tanks are made of steel, aluminum, and/or composite materials and can resist puncture much better than gasoline tanks.

Despite the overall safety of natural gas compared to other liquid fuels, there are a number of refueling safety concerns to be considered for both CNG an LNG. Users of CNG are concerned about failures of the pressure relief system (see Box 1-1, below) and leakage. In an enclosed area, released natural gas will rise to the ceiling, and in the absence of proper ventilation systems the build-up could result in the risk of a fire or explosion. The built up natural gas could also cause asphyxiation, if enough oxygen is displaced. In addition, if high-pressure refueling systems fail, injury could result from release of small particles, fire, or excessive noise. The primary hazard of LNG is frostbite due to direct skin exposure. Fire and explosion hazards of released LNG are similar to those of CNG.

---

[15] The inside tank is surrounded first by a layer of insulation and then by an outside tank.

> **Box 1-1.   CNG Safety: Pressure Relief Systems**
>
> CNG storage tanks need pressure relief systems to prevent overpressure failure of the tanks, most likely due to the risk of being subjected to a fire from a vehicle collision. There are two types of pressure relief systems. The first ones developed are simple caps with a soft metal in the center (usually a solder alloy) that melts when subjected to a fire. When the metal melts, the contents of the tank are released. Today, such pressure relief devices may only be used on Type I and II tanks (all metal and metal with composite reinforcement, respectively). For Type III and IV tanks (composite with metal liner and composite with plastic liner, respectively) the pressure relief system must use a mechanical device to relieve the pressure, most often a spring-loaded valve of some sort (these are commonly called "pressure relief devices," or PRDs). The reason that the Type III and IV tanks must use a PRD is because PRDs can be designed for higher flow rates which are needed for these tanks because they are weakened more quickly in fires than Type I and II tanks.
>
> The first wide-scale use of PRDs was in transit buses, where they earned a bad reputation for failures. These failures were due to a combination of a lack of design standards at the time, little experience in the field, insufficient testing, and a single supplier. Most of these failures occurred during refueling, and many transit bus PRDs were found to be releasing at pressures that were lower than the maximum pressures achieved during refueling. When a PRD releases at a refueling facility, the risk of fire should be low since refueling facilities should be built without presenting ignition sources. In addition, refueling stations have emergency shut-downs for such situations. The early problematic PRDs have all been replaced and PRD failures on transit buses is now a rare occurrence.
>
> PRDs are essential components of CNG tanks and provide a needed safety function. Failure of PRDs is rare, and they fail most often during refueling which should be an environment designed without ignition sources. Release of natural gas from the failure of a PRD presents hazards similar to most fuel spills.

1.6   Current Trends in Deployment of NGVs

As of August 2000, there were approximately 1.1 million converted NGVs in operation throughout the world.[16] In the United States alone, there were 103,673 converted and original equipment NGVs (which represents on the order of 0.05 percent of all vehicles), a 1.7 percent increase from 1999. Of the U.S. total, 101,991 vehicles ran on CNG and 1,682 vehicles ran on LNG, an increase from 1999 of 1.6 percent and 18.3 percent, respectively. Since 1992, the average annual growth rate in the U.S. for CNG and LNG vehicles were 20.3 percent and 44.2 percent, respectively.[17] Most NGVs in the United States and elsewhere are converted vehicles, although the number of dedicated vehicles offered is increasing.[18]

As illustrated in Figure 1-1, Argentina and Italy together lead the world with 462,186 and 320,000 converted NGVs, making up more than half the world total.[19] The United States, Brazil, and Russia follow these two countries, with approximately 104,000, 60,000, and 30,000 vehicles, respectively.

---

[16] International Association for Natural Gas Vehicles, "International Natural Gas Vehicle Statistics 2000 Online," http://www.iangv.org/html/ngv/stats.html.

[17] Energy Information Administration, "Alternatives to Traditional Transportation Fuels 1998," Table 1, http://www.eia.doe.gov/cneaf/solar.renewables/alt_trans_fuel98/table1.html.

[18] The U.S. Department of Energy Clean Cities Program defines a converted vehicle as a vehicle that was originally designed to operate on gasoline but has been altered to run on alternative fuel. The two most common fuel-switching alternatives include CNG and liquefied petroleum gas (LPG), also commonly referred to as propane.

[19] Similar statistics on OEM NGVs are not readily available.

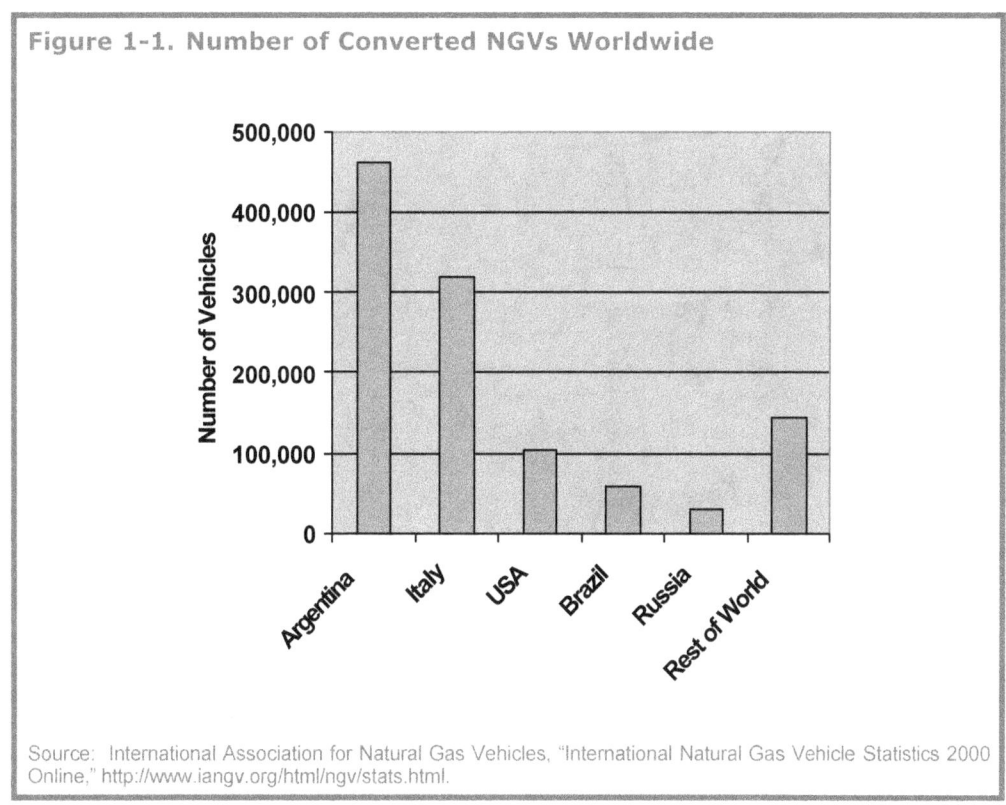

Figure 1-1. Number of Converted NGVs Worldwide

Source: International Association for Natural Gas Vehicles, "International Natural Gas Vehicle Statistics 2000 Online," http://www.iangv.org/html/ngv/stats.html.

Despite trailing other countries in numbers of NGVs, the United States has the highest number of NGV refueling stations. The total number of NGV refueling stations worldwide is 3,885, with approximately 32 percent (1,263) of these located in the United States, followed by Argentina (830), Italy (320), Canada (222), and Venezuela (151). As for individual States, California leads the nation with more than 200 refueling stations followed by Texas (77), Georgia (69), Utah (63), and New York (59). Only four states, Alaska, Hawaii, Maine, and Vermont do not have any NGV refueling sites (see Figures 1-2 and 1-3). However, the number of U.S.-based NGV refueling stations still pales in comparison to the approximately 180,000 gasoline stations throughout the United States.

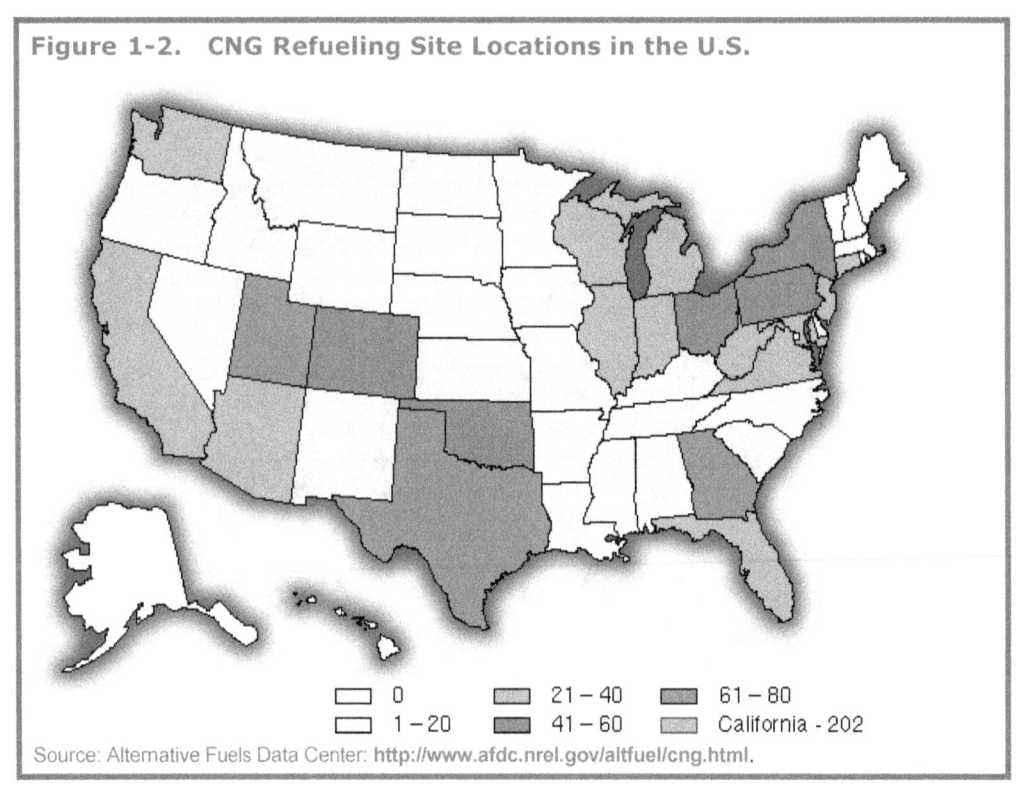

Figure 1-2. CNG Refueling Site Locations in the U.S.

Legend:
- ☐ 0
- ☐ 1 – 20
- ▨ 21 – 40
- ▨ 41 – 60
- ▨ 61 – 80
- ▨ California - 202

Source: Alternative Fuels Data Center: http://www.afdc.nrel.gov/altfuel/cng.html.

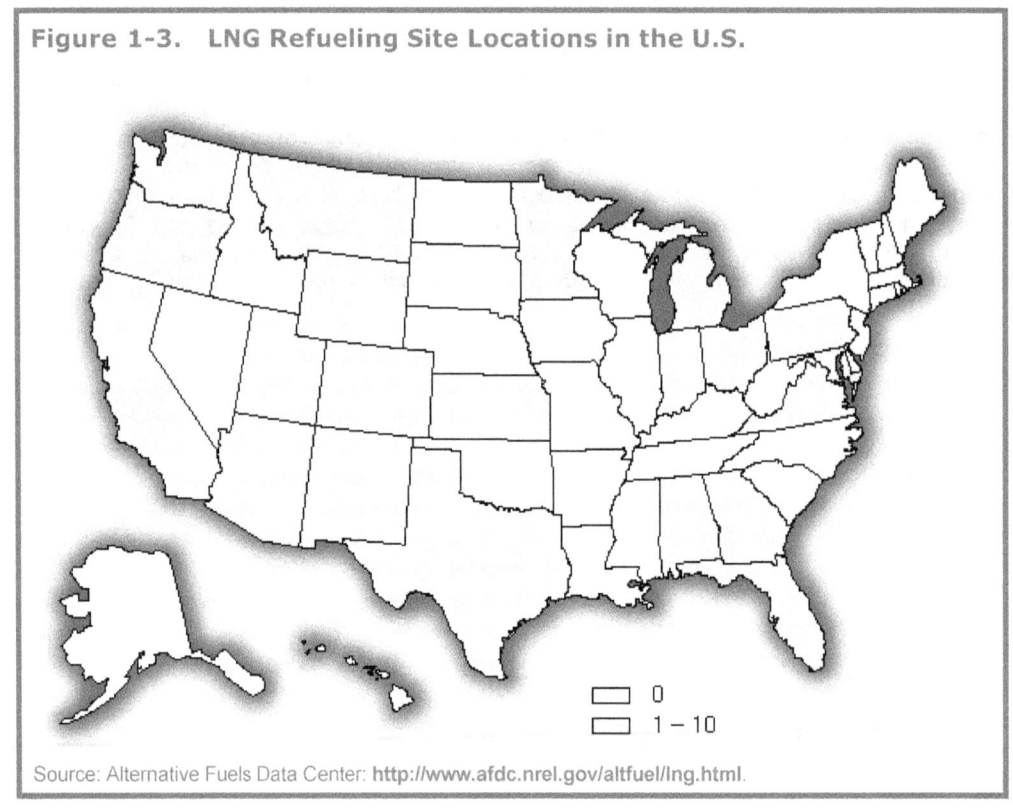

Figure 1-3. LNG Refueling Site Locations in the U.S.

Legend:
- ☐ 0
- ☐ 1 – 10

Source: Alternative Fuels Data Center: http://www.afdc.nrel.gov/altfuel/lng.html.

1 Natural Gas Vehicle Technology Options

2 Regulatory and Policy Frameworks Promoting Natural Gas Vehicles

Numerous regulatory policies have been introduced in the U.S. and abroad to promote the use of NGVs and facilitate the development of NGV projects along with a range of other AVFs.[20] Many of these policies are intended to help improve urban air quality and reduce dependence on fossil fuels, and some offer the indirect added benefit of reducing GHG emissions. This chapter considers a number of relevant regulations, policies, and programs that encourage the adoption of NGVs, as well as the development and implementation of new NGV technologies that can be used in NGV projects. The chapter provides an overview of the following: (1) federal laws and regulations affecting NGVs; (2) state initiatives; (3) voluntary programs and support activities promoting NGVs; and (4) international climate change programs.

2.1 Federal Laws and Regulations

Many of the most relevant elements of Federal policy to promote the development and use of alternative fuels in the transportation sector were introduced with the passage of the Energy Policy Act of 1992 (EPAct).[21] The primary motivations behind promoting alternative fuels under EPAct included reducing the nation's dependence on foreign oil and increasing the nation's energy security through the use of domestically produced alternative fuels.[22] To do so, EPAct established a goal of replacing 10 percent of petroleum-based motor fuels in the United States by the year 2000 and 30 percent by the year 2010. The statute also adopts the goal of seeking to reduce air, water, and other environmental impacts—including emissions of greenhouse gases—that result from the combustion of fossil fuel through transportation and other energy-consuming activities.[23] As discussed below, EPAct addresses NGVs in two principal ways: first, by providing tax credits and deductions for the purchase of AFVs and development of AFV infrastructure, and second, by mandating Federal, State, and private "alternative fuel provider" fleets to purchase AFVs.[24]

---

[20] The term "alternative fueled vehicle" is defined as any dedicated vehicle or a dual fueled vehicle. (Energy Policy Act of 1992) As provided in EPAct, the term "alternative fuel" is defined as: methanol, denatured ethanol, and other alcohols; mixtures containing 85 percent or more (or such other percentage, but not less than 70 percent, as determined by the Secretary, by rule, to provide for requirements relating to cold start, safety, or vehicle functions) by volume of methanol, denatured ethanol, and other alcohols with gasoline or other fuels; natural gas; liquefied petroleum gas; hydrogen; coal-derived liquid fuels; fuels (other than alcohol) derived from biological materials; electricity (including electricity from solar energy); and any other fuel the Secretary determines, by rule, is substantially not petroleum and would yield substantial energy security benefits and substantial environmental benefits.

[21] Energy Policy Act of 1992, Public Law 102-486.

[22] EPAct §2001.

[23] Id.

[24] According to the U.S. Department of Energy, an "alternative fuel provider" is defined as: [an entity] that owns, operates, leases, or otherwise controls 50 or more light-duty vehicles (LDVs) in the U.S. that are not on the list of EPAct Excluded Vehicles [such as emergency or law enforcement vehicles]: at least 20 of those LDVs are used primarily within a Metropolitan Statistical

2.1.1 Federal Tax Incentives for Natural Gas Vehicles

The Federal government introduced two forms of tax incentives relating to NGVs under EPAct: (1) a Federal tax deduction available to individuals and businesses purchasing qualified clean-fuel vehicles; and (2) a Federal tax deduction for business expenses related to the incremental cost to purchase or convert to qualified clean fuel vehicles.

Clean Fuel Vehicle Deduction

Title XIX of EPAct amended the Internal Revenue Code to provide tax deductions for the purchase of clean-fuel vehicles and certain refueling property, or for the conversion of a vehicle into one using clean-burning fuel.[25] Under those provisions, a qualified clean-fuel vehicle is one that operates using a "clean-burning fuel", including natural gas and liquefied natural gas, among other alternative fuels.[26] These provisions make available a Federal income tax deduction of up to between $2,000 and $50,000 per vehicle for the incremental cost to purchase or convert gasoline-powered vehicles into qualified clean fuel vehicles—including NGVs—and a deduction of up to $100,000 for certain kinds of property used for refueling these vehicles (see below).[27]

The deductions are available for clean fuel vehicles put into service between December 20, 1993 and December 31, 2004.[28] After an introductory period for the deduction ending in 2001, the deduction amount is reduced by 25 percent of the original amount each year after 2001, and will be phased out completely by 2005. The tax deduction for clean fuel vehicles is available for any applicable business or personal vehicle, except for certain electric vehicles that are eligible for a separate tax credit under related provisions. The deduction is not amortized and must be taken in the year the vehicle is acquired.[29]

As provided in Table 2-1, the maximum tax deduction for trucks or vans with gross vehicle weight of between 10,000 and 26,000 lbs is $5,000 per vehicle. The maximum deduction is $50,000 per vehicle for trucks and vans over 26,000 lbs., or buses with seating capacity of 20 or more adults. Other clean fuel vehicles may qualify for up to a $2,000 deduction. Table 2-1 also provides the maximum deductions for vehicles put into service after 2001 and through 2004, the final year the deduction may be taken before it is fully phased out.

Area (MSA)/Consolidated Metropolitan Statistical Area (CMSA); those same 20 LDVs are centrally fueled or capable of being centrally fueled. LDVs are centrally fueled if they capable of being refueled at least 75% of the time at a location that is owned, operated, or controlled by any fleet, or under contract with that fleet for refueling purposes. An alternative fuel provider is covered under EPAct if its principal business involves one of the following: producing, storing, refining, processing, transporting, distributing, importing, or selling any alternative fuel (other than electricity) at wholesale or retail; generating, transmitting, importing, or selling electricity at wholesale or retail; or produces and/or imports an average of 50,000 barrels per day or more of petroleum, as well as 30% or more of its gross annual revenues are derived from producing alternative fuels. http://www.ott.doe.gov/epact/alt_fuel_prov.shtml.

[25] See Internal Revenue Code, 26 U.S.C. §179A. See also EPAct, Title XIX, Subtitle A - Energy Conservation 7 [sic] Production Incentives. See also IRS Publication 535 (2001), page 44.

[26] 26 U.S.C. §179A(e). Clean-burning fuels include: natural gas, liquefied natural gas, liquefied petroleum gas, hydrogen, electricity, and any other fuel at least 85 percent of which is 1 or more of the following: methanol, ethanol, any other alcohol, or ether. Id.

[27] 26 U.S.C. §179A(c).

[28] 26 U.S.C. §179A(g).

[29] U.S. Department of Energy, "Alternative Fuel Vehicle Fleet Buyer's Guide," http://www.fleets.doe.gov/cgi-bin/fleet/main.cgi?17357,state_ins_rep,5,468050; see also IRS Publication 535 (U.S. Department of Energy, 2001), page 44.

Table 2-1. Summary of Deductions for Clean Fuel Vehicles

Date Vehicle Acquired	Vehicle Type	Deduction Available
Dec. 20, 1993 - 2001	truck or van with GVW 10,000-26,000 lbs.	$5,000
	truck or van with GVW over 26,000 lbs.	$50,000
	each bus, with seating capacity of at least 20 adults (excluding driver)	$50,000
	all other vehicles (excluding off-road vehicles)	$2,000
2002	truck or van with GVW 10,000-26,000 lbs.	$3,750
	truck or van with GVW over 26,000 lbs.	$37,500
	each bus, with seating capacity of at least 20 adults (excluding driver)	$37,500
	all other vehicles (excluding off-road vehicles)	$1,500
2003	truck or van with GVW 10,000-26,000 lbs.	$2,500
	truck or van with GVW over 26,000 lbs.	$25,000
	each bus, with seating capacity of at least 20 adults (excluding the driver)	$25,000
	all other vehicles (excluding off-road vehicles)	$1,000
2004	truck or van with GVW 10,000-26,000 lbs.	$1,250
	truck or van with GVW over 26,000 lbs.	$12,500
	each bus, with seating capacity of at least 20 adults (excluding the driver)	$12,500
	all other vehicles (excluding off-road vehicles)	$500
2005	Deduction fully phased out for all vehicles	None

Source: 26 U.S.C. §§179A(b)-179A(c).

Deduction for Qualified Clean-Fuel Vehicle Refueling Property

Qualified clean-fuel vehicle refueling property is defined as property that is used "for the storage or dispensing of a clean-burning fuel" for use in a qualified clean-fuel vehicle.[30] The tax deduction available under Section 179A for such refueling property for each location where it is put into service is up to $100,000, minus the total deductions on all such property placed in service at the location in all earlier years.[31] The deduction for the property is not reduced in value over time, as it is for the qualified clean-fuel vehicles, but the deduction will end for clean-fuel vehicles put into service starting in 2005.[32]

Under EPAct, "alternative" fuels include natural gas, methanol, ethanol, propane, hydrogen, coal-derived liquids, biological materials, and electricity. EPAct also includes any other fuel that the Secretary of Energy finds to be substantially non-petroleum and which would yield substantial energy security and environmental benefits.

2.1.2 AFV Acquisition Requirements for Federal, State, and Alternative Fuel Provider Fleets

In addition to providing tax incentives for AFVs, EPAct created new AFV procurement mandates for Federal, state, and "alternative fuel provider" fleets to purchase AFVs, a large portion of which have been NGVs. EPAct first introduces AFV acquisition

---

[30] 26 U.S.C. §179A(d).

[31] See IRS Publication 535 (2001), page 46.

[32] For more information, contact Winston Douglas, Alternative Fuels Tax Provisions, at (202) 622-3110, fax (202) 622-4779; or Frank Boland, Alcohol Fuel Tax Information, at (202) 622-3130; or call the toll-free order desk at (800) 829-3676. U.S. Department of Energy, "Alternative Fuel Vehicle Fleet Buyer's Guide" http://www.fleets.doe.gov/cgi-bin/fleet/main.cgi?17357,state_ins_rep,5,468050.

requirements in Federal fleets, and these provisions have been underscored by several Executive Orders that further the commitments of Federal agency fleets to adopt AFVs. Likewise, state and alternative fuel provider fleets must meet the requirements outlined in the Alternative Fuel Transportation Program, Final Rule, under the EPAct implementing regulations.[33] At the time of publication of this report, the U.S. Department of Energy, in implementing EPAct, is considering whether to also extend EPAct's AFV procurement requirements to local government and private fleets, authorized under EPAct sections 507(g) and 507(k).[34] Federal, state and alternative fuel provider, and local and private AFV requirements are discussed as follows.[35]

EPAct Procurement Requirements for AFVs in Federal Fleets

Section 303 of EPAct requires the entire Federal government, under the direction of the Department of Energy, to acquire at least 5,000 light-duty AFVs in FY1993; 7,500 light-duty AFVs in FY1994; and 10,000 light-duty AFVs in FY1995. Following FY1995, all Federal fleets consisting of at least 20 or more light-duty motor vehicles operating in a "metropolitan statistical area"[36] must meet a specific percentage requirement for AFVs. These requirements include: 25 percent in FY1996; 33 percent in FY1997; 50 percent in FY1998; and 75 percent in FY1999 and thereafter.[37] These requirements are summarized in Table 2-2 below. (See "Success of the EPAct AFV Program for Federal Fleets" later in this section.)

Table 2-2. Summary of EPAct Requirements for Federal Government Acquisition of AFVs

Fiscal Year Vehicle Acquired	Applicable Fleet	Number of AFVs Required
FY1993		5,000 total light-duty AFVs
FY1994	Entire Federal Government	7,500 total light-duty AFVs
FY1995		10,000 total light-duty AFVs
FY1996		20% of each fleet as AFVs
FY1997	Each Federal fleet with 20 or more light-duty vehicles in a "metropolitan statistical area"	33% of each fleet as AFVs
FY1998		50% of each fleet as AFVs
FY1999 and thereafter		75% of each fleet as AFVs

Table 2-3 also summarizes the annual purchase requirements for Federal and State fleets, alternate fuel providers, and private and municipal fleets. Each of these fleets is described in the sections that follow.

---

[33] 10 CFR Part 490.

[34] See http://www.ott.doe.gov/epact/private_fleets.shtml.

[35] Additional information about the U.S. Department of Energy's AFV programs under EPAct, see http://www.ott.doe.gov/epact.

[36] EPAct §303 defines a "metropolitan statistical area" as having a population of 250,000 or more in 1980 according to the U.S. Census. This definition is not always consistent with other provisions of EPAct.

[37] EPAct §303.

		Table 2-3.	Summary of AFV Purchase Requirements under EPAct		
Model Year	Federal	State	AFV Provider	Private Fleets
---	---	---	---	---
1997	33%	10%	30%	0
1998	50%	15%	50%	0
1999	75%	25%	70%	0
2000	75%	50%	90%	0
2001	75%	75%	90%	0
2002	75%	75%	90%	20%
2003	75%	75%	90%	40%
2004	75%	75%	90%	60%
2005 and later	75%	75%	90%	70%

Source: DOE, Office of Transportation Technologies, "EPACT/Clean Fuel Fleet Program Fact Sheet," http://www.afdc.doe.gov/pdfs/caaa.pdf.

To encourage and promote the use of AFVs in Federal fleets, EPAct also creates an agency incentive program and a recognition and incentive awards program for Federal agencies. Under the Act, the General Services Administration (GSA) may offer a reduction in fees charged to agencies to lease AFVs below those fees charged for the lease of comparable conventionally fueled motor vehicles.[38] The GSA is also required to establish an annual awards program that recognizes Federal employees who have demonstrated "the strongest commitment to the use of alternative fuels and fuel conservation in Federal motor vehicles."[39] Moreover, the Act requires the U.S. Postal Service to provide a report to Congress outlining its AFV program.[40]

Executive Order 13149: Fuel Economy and AFV Procurement Requirements for Federal Fleets

Federal agencies have been required to follow guidelines established by several Executive Orders, starting with Executive Order 12844 (April 21, 1993) and Executive Order 13031 (December 13, 1996) that each underscored the policies and objectives of the Federal agency AFV provisions of EPAct. Both of those Orders were superceded by Executive Order 13149, signed in April 21, 2000, which further strengthened the Federal government's commitment to promote the use of all types of AFVs in Federal fleets.

Executive Order (E.O.) 13149 requires Federal agencies operating 20 or more motor vehicles within the United States to reduce the fleet's annual petroleum consumption by 20 percent below FY1999 levels by the end of FY2005.[41] To meet this goal, Federal agencies are given significant flexibility in developing an appropriate strategy to meet the petroleum reduction levels. Agencies are required to use alternative fuels, such as natural gas, to meet the majority of the fuel requirements for vehicle fleets operating in "metropolitan statistical areas," defined in E.O. 13149 as metropolitan areas with populations of more than 250,000 in 1995 according to the Census Bureau. Where feasible, the Order also instructs agencies to consider procuring "innovative" alternative fuel vehicles that are capable of large improvements in fuel economy, such as NGVs. Agencies are required to increase the average EPA fuel economy rating of their light-duty

---

[38] EPAct §306.
[39] EPAct §307.
[40] EPAct §311.
[41] E.O. 13149 §201. Independent agencies are encouraged but not required to comply with the Order. E.O. 13149 §504.

vehicle acquisitions by at least one mile per gallon (mpg) by 2002 and 3 mpg by 2005 above 1999 acquisition levels. Agencies are also encouraged to adopt awards and performance evaluation programs that reward federal employees for exceptional performance in implementing the Order.[42] Federal fleet requirements under E.O. 13149 are summarized in Table 2-4.

Table 2-4. Summary of E.O. 13149 Requirements for Federal Fleets

Applicable Fleet	Effective Date	Action Required
Each Federal fleet with 20 or more light-duty vehicles	FY2002	Increase average EPA fuel economy rating of light-duty vehicle acquisitions by 1 mpg above FY1999 levels
Each Federal fleet with 20 or more light-duty vehicles	FY2005	Increase average EPA fuel economy rating of light-duty vehicle acquisitions by 3 mpg above FY1999 levels
Each Federal fleet with 20 or more light-duty vehicles	By end of FY2005	Reduce fleet's annual petroleum consumption by 20% below FY1999 levels
Each Federal fleet with 20 or more light-duty vehicles operating in metropolitan statistical areas	By end of FY2005	Same action as above, but must include alternative fuels to meet majority of fuel requirements

E.O. 13149 also establishes an AFV acquisition credit program for Federal agencies pursuant to the requirements under EPAct. In preparing an annual report to DOE and the Office of Management and Budget (OMB), each Federal agency acquisition of a light-duty AFV counts as one credit towards fulfilling EPAct's AFV acquisition requirements. Agencies receive one additional credit for each light-duty AFV that exclusively uses an alternative fuel, and for each zero emission vehicle. Agencies receive three credits for dedicated medium-duty AFVs and four credits for dedicated heavy-duty AFVs.[43] Table 2-5 summarizes the number of credits available for each type of acquired AFV.

Table 2-5. Summary of Credits for Federal Fleet Acquisitions of AFVs under Executive Order 13149

Type of AFV	Number of Credits Awarded
Each light-duty AFV	1 credit
Each light-duty AFV exclusively using an alternative fuel	2 credits
Each ZEV	2 credits
Each dedicated medium-duty AFV	3 credits
Each dedicated heavy-duty AFV	4 credits

Fleet owners may use these credits to meet acquisition requirements in later years or to sell and trade credits with other fleets. Thus, fleet owners that do not meet the E.O. acquisition requirements for AFVs may purchase credits from fleet owners with a surplus of AFVs credits.

In order to provide for adequate access to refueling infrastructure, Federal agencies are directed under E.O. 13149 to "team with state, local, and private entities to support the expansion and use of" public refueling stations for AFVs.[44] State, local, and private groups may also establish non-public alternative fuel stations if no commercial infrastructure is available in their territory.[45]

---

[42] E.O. 13149 §303
[43] E.O. 13149 §401.
[44] E.O. 13149 §402(a).
[45] E.O. 13149 §402(b).

According to the Department of Energy (DOE) Clean Cities Report *Federal Fleet AFV Program Status,* dated June 2, 1998, as of 1998, of more than 570,000 vehicle acquisitions overall, the estimated cumulative total AFV acquisitions in Federal agencies totaled more than 34,000 vehicles between FY1991 and FY1998. This represented about 80 percent compliance with the 44,600 required AFV acquisitions under EPAct. Of those AFVs acquired by 1998, approximately 18,000, or 52 percent, were CNG vehicles. The majority of those vehicles were converted from existing gasoline vehicles, and were predominantly performed on U.S. Postal Service and Department of Defense vehicles.[46]

As a result of the missed target for Federal AFV acquisitions under EPact, in January 2002 three environmental organizations filed a lawsuit in Federal court against 17 Federal agencies for failing to comply with EPAct.[47] The plaintiffs claim that all 17 agencies have failed: (1) to meet their AFV acquisition requirements; (2) to file the necessary compliance reports with Congress; and (3) to make these reports available to the public. The complaint also alleges that DOE failed to complete a required private and municipal AFV fleet rulemaking. As a remedy, the plaintiffs request that the court order the agencies to comply with these requirements, and to require the agencies to offset their future vehicle purchases with the number of AFVs necessary to bring them into compliance with EPAct's acquisition requirements for 1996 through 2001. A decision on the case is pending.[48]

EPAct Procurement Requirements and Incentives for AFVs in Alternative Fuel Provider and State Fleets

In 1996, DOE issued final regulations that spell out fleet responsibilities under the State and Alternative Fuel Provider Program. Like the Federal fleet requirements, this is a DOE regulatory program that requires covered state and "alternative fuel provider" fleets to purchase AFVs as a portion of their annual light-duty vehicle acquisitions.[49]

As required for Federal fleets, EPAct requires Alternative Fuel Providers to acquire AFVs as a portion of their annual light-duty acquisitions, starting with Model Year (MY) 1996.[50] The implementing regulations under EPAct Section 501 provide a schedule for *alternative fuel providers* to acquire light-duty AFVs as follows: 30 percent for model year 1997; 50

---

[46] U.S. Department of Energy, Federal Fleet AFV Program Status (June 2, 1998), available at: http://www.ccities.doe.gov/pdfs/slezak.pdf. As stated in the report:

> Of the 34,000+ AFVs acquired by Federal agencies, approximately 10,000 (30 percent) have been M-85 (methanol mixed with gasoline) flexible fuel vehicles, 6,000 (17 percent) have been E-85 (ethanol mixed with gasoline) flexible fuel vehicles, and 18,000 (52 percent) have been compressed natural gas (CNG) vehicles. Several hundred each of electric and liquefied petroleum gas (LPG or propane) vehicles have also been acquired. Projections for future Federal AFV acquisitions, based on discussions with Federal agencies' procurement personnel and manufacturers, indicate that flexible fuel E-85 vehicles will be the most common AFV procured by agencies' to comply with EPACT, followed by CNG.

[47] Center for Biological Diversity v. Abraham, N.D. Cal., No. CV-00027 (January 2, 2002). The agencies named in the suit include: the Departments of Energy, Commerce, Justice, Interior, Veterans Affairs, Agriculture, Transportation, Health and Human Services, Housing and Urban Development, Labor, State, and Treasury; the Environmental Protection Agency; the U.S. Postal Service; the National Aeronautics and Space Administration; the U.S. Nuclear Regulatory Commission; and the General Services Administration.

[48] See http://www.evaa.org.

[49] EPAct §501; 10 CFR 490.303.

[50] EPAct §501. See generally 10 CFR 490.

percent for model year 1998; 70 percent for model year 1999; and 90 percent for model year 2000 and thereafter.[51]   (See Table 2-6.)

Table 2-6.	EPAct Requirements for Light-duty AFV Acquisitions for Alternative Fuel Providers
Model Year Vehicle Acquired	**Percentage of AFVs Required**
MY1997	30 percent
MY1998	50 percent
MY1999	70 percent
MY2000 and thereafter	90 percent

The AFV regulations cover a state agency if it owns or operates 50 or more light-duty vehicles, at least 20 of which are used primarily within a metropolitan area.[52]   States are required to prepare plans for implementing an AFV program and various policy incentives that may be used to encourage the adoption of AFVs.[53]   The mandatory acquisition schedule of AFVs for *state government* fleets is: 10 percent for model year 1997; 15 percent for model year 1998; 25 percent for model year 1999; 50 percent for model year 2000; and 75 percent for model year 2001 and thereafter.[54]   (See Table 2-7.)

Table 2-7.	EPAct Requirements for Light-duty AFV Acquisition for State Fleets
Model Year Vehicle Acquired	**Percentage of AFVs Required**
MY1997	10 percent
MY1998	15 percent
MY1999	25 percent
MY2000	50 percent
MY2001 and thereafter	75 percent

Like the Federal program, Alternative Fuel Providers and state fleets earn one credit for every light-duty AFV acquired every year above the base AFV acquisition requirements. Once they have satisfied their annual light-duty AFV acquisition requirements, covered fleets may also earn one credit for every heavy-duty AFV acquired annually. Again, these credits are freely transferable between fleets, or can be banked for future years. DOE has created a Credit Trades Bulletin Board to assist fleets in buying or selling AFV credits.[55]

Other EPAct Incentives for AFVs

The following additional provisions may encourage the use of alternative fuels:

- Up to $30 million/year to assist in the purchase of alternate fuel transit buses and school buses;

---

[51] 10 CFR 490.302.

[52] See *Federal Register*, Volume 61, Number 51, pages 10627-10628.

[53] EPAct §409.

[54] EPAct §507(o); 10 CFR 490.201.

[55] EPAct 508(d); 10 CFR 409. See also Alternative Fuel Transportation Program, Final Rule, 10 CFR Part 490), http://www.fleets.doe.gov/cgi-bin/fleet/main.cgi?17357,state_ins_rep,5,468050. See also http://www.ott.doe.gov/epact/state_fleets.shtml for more information.

- $25 million/year for low-interest loans for the purchase of AFVs;
- State and local incentive programs, including $10 million/year to assist states in acquiring AFVs;
- Exemption for vehicular natural gas from certain Federal and State regulations;
- Certification of training programs for alternate fuel vehicle technicians; and
- Public information programs.

Success of the EPAct AFV Program for Alternative Fuel Vehicle and State Fleets

According to the 2001 annual report for the State & Alternative Fuel Provider (S&FP) program, covered fleets were required to purchase a total of 13,501 light-duty AFVs in MY2000. The fleets slightly exceeded this number, purchasing a total of 13,541 light-duty AFVs. In addition, fleets banked an excess of 4,101 credits during MY2000. As shown in Figure 2-1, dedicate and bi-fuel compressed natural gas vehicles were strongly favored by S&FP fleets.

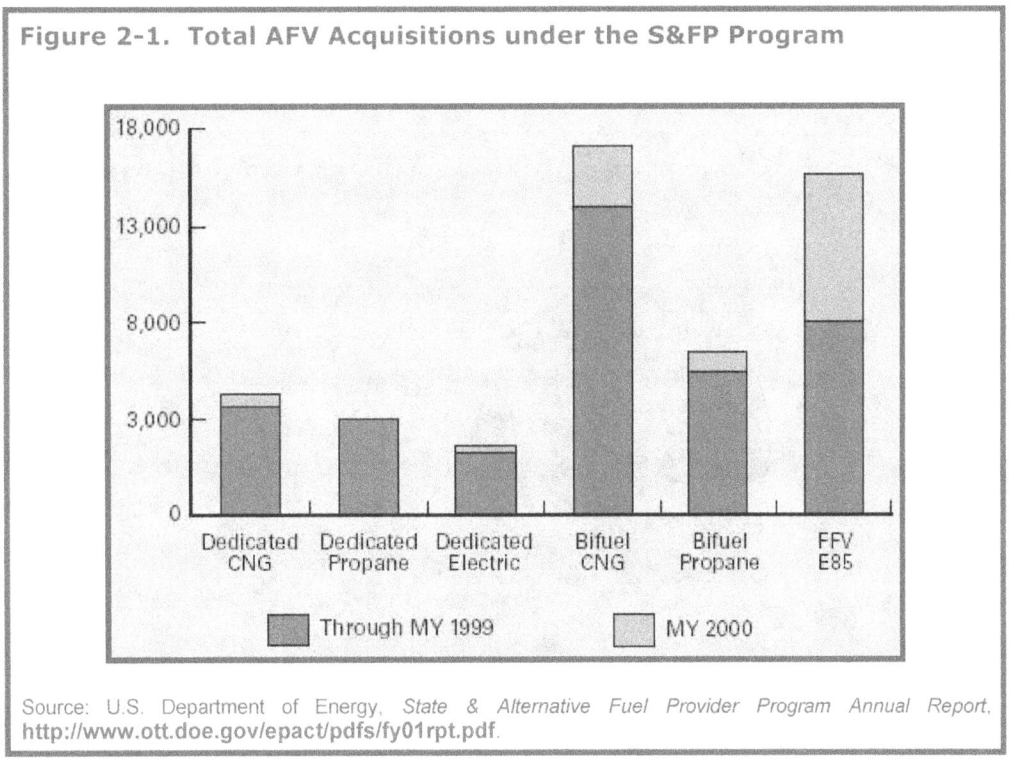

Figure 2-1. Total AFV Acquisitions under the S&FP Program

Source: U.S. Department of Energy, *State & Alternative Fuel Provider Program Annual Report*, http://www.ott.doe.gov/epact/pdfs/fy01rpt.pdf.

For MY2000, a total of 376 credits were traded by 12 fleets, accounting for less than 2% of the total credit activity for MY 2001. In combination with the fact that the total number of credits banked by fleets remains at the high level of 46,155, this suggests most fleets are saving credits for their own use. During MY 2000, fleets used 2,759 banked credits towards meeting their compliance requirements.[56]

---

[56] U.S. Department of Energy Office of Transportation Technologies, "Program Activity and Accomplishments in FY2001," (Washington, D.C., December 2001), http://www.ott.doe.gov/epact/pdfs/fy01rpt.pdf.

According to the 2001 Annual Report, only about 9% of the S&FP fleets had failed to comply with program requirements[57] Figures 2-2 and 2-3 indicate overall compliance and vehicle acquisition trends in state and alternative fuel provider fleets.

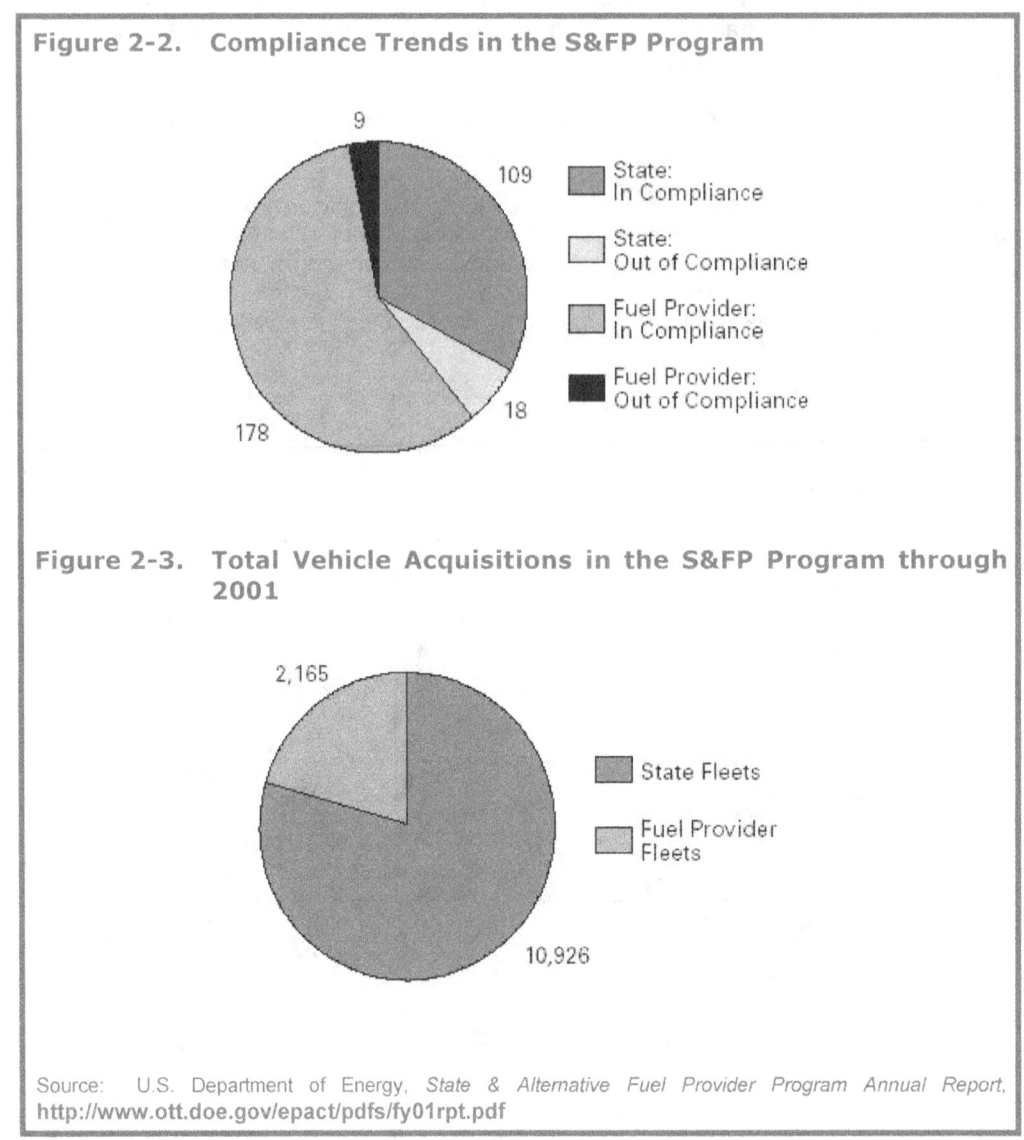

Figure 2-2.   Compliance Trends in the S&FP Program

- State: In Compliance
- State: Out of Compliance
- Fuel Provider: In Compliance
- Fuel Provider: Out of Compliance

Figure 2-3.   Total Vehicle Acquisitions in the S&FP Program through 2001

- State Fleets
- Fuel Provider Fleets

Source: U.S. Department of Energy, *State & Alternative Fuel Provider Program Annual Report*, http://www.ott.doe.gov/epact/pdfs/fy01rpt.pdf

Preliminary results of MY2001 acquisitions reporting in the Spring 2002 indicated that states and alternative fuel provider fleets collectively acquired more than 60,000 AFVs since the launch of the program, again exceeding the program quota.[58]

---

[57] Id.
[58] U.S. Department of Energy Office of Transportation Technologies, "What's New: Spring 2002 Update," (Washington, D.C., May 2002), http://www.ott.doe.gov/epact/pdfs/whatsnew_spring_02.pdf.

Pursuant to EPAct Section 507(g), the Department of Energy is currently considering whether to adopt and implement an AFV acquisition program for other fleets, i.e., local government and private fleets.[59] Before implementing a potential Private and Local Government (P&LG) fleet program, DOE must also determine whether doing so would be necessary to help meet the EPAct's U.S. petroleum replacement fuel goals, and that it is technically and economically practical.[60] The DOE may also consider whether to include law enforcement motor vehicles and new urban buses as part of the program.[61]

Under such a prospective program, local governments or private fleets would be covered if they own or operate at least 50 light-duty vehicles in the U.S., 20 of which are primarily used within a metropolitan statistical area.[62] EPAct outlines the percentage of AFVs that would have to be acquired for each model year, should DOE adopt such a program, as sown in Table 2-8.

Table 2-8. EPAct Requirements for Light-duty AFV Acquisition for All Other Fleets	
Model Year Vehicle Acquired	**Percentage of AFVs Required**
MY1999, 2000, 2001	20 percent
MY2002	30 percent
MY2003	40 percent
MY2004	50 percent
MY2005	60 percent
MY2006 and thereafter	70 percent

2.1.3 Clean Air Act of 1970 and Clean Air Act Amendments of 1990

General Provisions

The Clean Air Act (CAA), first enacted in 1970 and amended in 1990, provides the basis for the Federal government's authority to address air pollution throughout the United States under the authority of the U.S. Environmental Protection Agency, including the regulation of emissions from stationary and mobile sources. Starting with Section 202 of the 1990 Amendments, the CAA establishes emission and fuel standards for mobile sources, and provides standards for clean fuel vehicles, including light-duty clean fuel vehicles, light-duty trucks, and flexible and dual-fuel vehicles.[63] The CAA also allows the State of California to promulgate its own standards for clean fuel vehicles.[64]

---

[59] EPAct §507(e).
[60] EPAct §507(a)(3). See DOE website at http://www.ott.doe.gov/epact/private_fleets.shtml.
[61] EPAct §507(k).
[62] EPAct §301.
[63] CAA §243. Under the Act, "clean fuels" are defined as natural gas, ethanol, methanol or other alcohols; mixtures containing 85 percent or more methanol, ethanol or other alcohols; reformulated gasoline and diesel; propane; electricity; and hydrogen. CAA §241.
[64] CAA §243.

Clean Fuel Fleet Program

The 1990 Amendments also established the Clean Fuel Fleet (CFF) Program that requires "covered fleets" with 10 or more vehicles owned by public or private entities in Consolidated Metropolitan Statistical Areas (CMSAs)[65] to acquire clean-fuel vehicles (CFVs) when replacing existing vehicles. Under the Act, states would have the option of adopting an alternative program under the state's State Implementation Plan under the CAA, so long as the state would meet the equivalent reductions in ambient emissions. To date, CMSAs that states have opted to include in the CFF Program include Atlanta, Georgia; Chicago-Gary-Lake County, Illinois/Indiana; Denver-Boulder, Colorado; and Milwaukee-Racine, Wisconsin.

As required under the CAA, starting in model year 1999, 30 percent of new light-duty vehicles were required to be clean fuel vehicles and 50 percent of newly acquired medium- and heavy-duty vehicles (i.e., 8,500 - 26,000 gross vehicle weight) were required. (Fleets composed of law enforcement and emergency vehicles are exempt from the requirements.) Required procurement levels increase in following years, as provided in Table 2-9.

Table 2-9. Purchasing Requirements under the Clean Fuel Fleet Program[66]			
Vehicle Size	1999	2000	2001 and later
GFW Rated less than 8,500 lbs	30%	50%	70%
GFW Rated less than 26,500 lbs	50%	50%	50%

The CFF Program offers credits for each clean fuel vehicle purchased under the program, based on the emission level of the vehicle. Low emission vehicles (LEVs) receive one credit, ultra-low emission vehicles (ULEVs) receive two credits, and zero emission vehicles (ZEVs) receive three credits each. Credits may be used to demonstrate compliance with the program, and may be freely traded to meet compliance requirements by participating fleets as needed.[67]

2.1.4 Federal "Inherently Low-Emission Airport Vehicle" (ILEAV) Pilot Project

In 2000, Congress passed the Wendell H. Ford Aviation Investment and Reform Act for the 21st Century which included provisions to establish a $20 million program to introduce low emission vehicles at 10 airports (identified by the Department of Transportation) that are located in air quality non-attainment areas as defined by the Clean Air Act.[68] Under the law, the Federal government commits 50 percent of the funding for the pilot projects to introduce natural gas and other clean fuel vehicles to airport fleets, as well as to implement clean fuel infrastructure.[69]

In May 2001, the Federal Aviation Administration (FAA) announced the 10 airports selected for the ILEAV program out of 40 that had expressed interest. The selected

---

[65] CAA §241(5). CMSAs include cities are metropolitan areas that had a population of at least 250,000 in 1980 and have been classified as extreme, severe, or serious non-attainment areas for ozone as defined by the CAA. At the time of the passage of the CAA Amendments in 1990, 22 metropolitan areas would have qualified. CAA §246.

[66] CAA §246. See also National Alternative Fuels Hotline, *The Clean Fuel Fleet Program* (September 1998), located at http://www.afdc.doe.gov/pdfs/caaa.pdf.

[67] CAA §246.

[68] Public Law No: 106-181, Section 133.

[69] Id.

airports include: Baltimore-Washington International (BWI); Baton Rouge, Louisiana; Metropolitan Chicago O'Hare International; Dallas/Fort Worth International; Denver International; Hartsfield Atlanta International; New York's John F. Kennedy International; New York LaGuardia; Sacramento International; and San Francisco International. Of these airports, BWI, Baton Rouge, and Denver will use compressed natural gas (CNG) vehicles only, and all others except Dallas/Ft. Worth and Chicago will use a combination of NGVs and EVs. Of the 2,000 vehicles funded in the program, about 35 percent will be CNG, although 540 out of the 600 curbside and on-road vehicles (i.e., other than tarmac vehicles) will use CNG.[70]

2.2   State Laws and Policies

A growing number of states have adopted policy measures promoting the use of NGVs. Some of the larger programs, such as in California and several northeastern states are discussed below.

2.2.1   California

Section 209(a) of the Clean Air Act prohibits states from adopting or enforcing standards for new motor vehicles or new motor vehicle engines—with the exception of the State of California.[71] In response to California's severe air pollution problems, CAA Section 209(b) grants the state the explicit authority to set its own standards for vehicular emissions, so long as the standards are equal to, or more stringent than, those set by the CAA and are approved by EPA.[72] State studies have found that about half of smog-forming pollutants are produced by gasoline and diesel-powered vehicles, and that only alternative technologies would help California reduce motor vehicle air pollution that will result from increasing driving rates in the State.[73]

California's response to its severe air quality problems was the adoption of a series of regulations in the 1990s to promote the adoption of new LEVs and ZEVs in the state. The Clean Air Act permits other states to follow California so long as any motor vehicle emissions regulations adopted by those states are identical to California's.[74] Since California introduced its LEV standards in 1990, four other States—New York, Massachusetts, Maine, and Vermont—have adopted the California emissions requirements for a percentage of motor vehicles sold in those states.

As of this time, a number of light-duty NGVs meet several of the strict emissions standards established under the LEV and ZEV programs. A list of these vehicles is provided in Table 2-13, below.

In addition to the LEV provisions, the State of California has adopted a number of policies and undertaken a wide range of programs promoting AFVs in the state. Natural gas is the leading alternative fuel in use in California in terms of the number of LEVs and fueling stations available, and it appears that compressed natural gas and liquefied natural gas will be widely employed to meet additional emission regulations.[75]

---

[70] Natural Gas Vehicle Association, Airports & AFVs, see website http://www.ngvc.org/ngv/ngvc.nsf/bytitle/airportsandafvs.html.

[71] 42 U.S.C. 7609(a).

[72] 42 U.S.C. 7609(b).

[73] California Air Resources Board, California's Zero Emission Vehicle Program, "Fact Sheet," California Air Resources Board (December 26, 2001), http://www.arb.ca.gov/msprog/zevprog/factsheet/evfacts.pdf.

[74] 42 U.S.C. 7507.

[75] California Energy Commission, "California Clean Fuel Market Assessment 2001," P600-01-018 (September 2001), pages 12-13.

Low Emission Vehicle (LEV) Regulatory Program

The flexibility provided to California under the CAA paved the way for sweeping regulation that has established extensive standards for low and zero emissions vehicles sold in the State. In 1990 the California Air Resources Board (CARB) adopted the first set of regulations to require automobile manufacturers to introduce LEVs to the California automobile market. The regulations require manufacturers to sell a certain percentage of these vehicles each year. Known as LEV I, the new standards promised to affect the entire automobile market in California by introducing various new LEVs, including a number of NGVs with reduced emissions.

LEV I standards were based on the introduction of four classes of vehicles with increasingly more stringent emissions requirements. Under the LEV I requirements, as of 1994 manufacturers were permitted to certify vehicles in any combination of the LEV categories through 2003 in order to satisfy the LEV standard.[76] It should be noted that under current regulations, auto manufacturers are also required to comply with a fleet-based average Non-Methane Organic Gas standard (NMOG), which introduces more and more stringent standards with each model year.[77]

Following a hearing in November 1998, the CARB amended the LEV I regulations and adopted LEV II, the second-generation LEV program. While the first set of LEV standards covered 1994 through 2003 models years, the LEV II regulations cover 2004 through 2010 and represent continued emissions reductions. The LEV II amendments were formally adopted by the CARB on August 5, 1999 and came into effect on November 27, 1999.[78]

Under LEV II, manufacturers may certify vehicles under one of five emission standards, listed in order from least to most stringent:

- transitional low emissions vehicles (TLEVs)
- low-emission vehicles (LEVs);
- ultra-low-emission vehicles (ULEVs);
- super ultra-low emissions vehicles (SULEVs); and
- zero emissions vehicles (ZEVs).

The more stringent LEV II regulations were adopted in part to keep up with changing passenger vehicle fleets in the state, where more sport utility vehicles (SUVs) and pickup trucks are used as passenger cars rather than work vehicles. The LEV II standards were a necessary step for the state to meet the Federally-mandated CAA goals that address ambient air quality standards as outlined in the 1994 State Implementation Plan (SIP).[79] LEV II increased the stringency of the emission standards for all light- and medium-duty vehicles beginning with the 2004 model year and expanded the category of light-duty trucks up to 8,500 lbs. gross vehicle weight (including almost all SUVs) to be subject to

---

[76]See California Air Resources Board, "California Exhaust Emissions Standards and Test Procedures for 2001 and Subsequent Model Passenger Cars, Light-duty Trucks, and Medium-Duty Vehicles," Proposed Amendments (September 28, 2001).

[77] §1960.1(g)(2). California's fleet average NMOG mechanism "requires manufacturers to introduce an incrementally cleaner mix of Tier 1, TLEV, LEV, ULEV and ZEV vehicles each year, with the fleet average NMOG value for passenger cars and lighter light-duty trucks decreasing from 0.25 gram/mile in the 1994 model year to 0.062 gram/mile in the 2003 model year." See California Air Resources Board, "The California Low-Emission Vehicle Regulations" (May 30, 2001), http://www.arb.ca.gov/msprog/levprog/cleandoc/levregs053001.pdf.

[78] California Low-Emission Vehicle Program, see website http://www.arb.ca.gov/msprog/levprog/levprog.htm.

[79] California Low-Emission Vehicle Program, see website http://www.arb.ca.gov/msprog/levprog/levprog.htm.

the same standards as passenger cars.[80] When LEV II is fully implemented in 2010, it is estimated that smog-forming emissions in the Los Angeles area will be reduced by 57 tons per day, while the statewide reduction is expected to be 155 tons per day.[81]

The LEV II standards go further to require that vehicles classified as LEV and ULEV meet NO_x standards which are 75 percent below LEV I requirements based on fleet averages. In addition, fleet average durability standards are extended from 100,000 to 120,000 miles. LEV II also allows manufacturers to receive credits for vehicles meeting near-zero emissions, and a new category of vehicles called super ultra-low emissions vehicles (SULEVs).[82] The LEV II standards were also designed to respond to some delays and "inertia" the LEV program had been facing, and pushed back the starting year of the program to 2003.

Some examples of LEV I and LEV II emissions standards for the different vehicles types are provided in Tables 2-10 and 2-11.

Table 2-10.	LEV I Exhaust Emission Standards for New MY2001-MY2003 Passenger Cars and Light-duty Trucks (3,750 lbs. LVW or less)					
Durability of Vehicle	Vehicle Emission Category	NMOG (g/mi)	Carbon Monoxide (g/mi)	NOx (g/mi)	Formaldehyde (mg/mi)	Particulates fr. diesel vehicles (g/mi)
50,000	Tier 1	0.250	3.4	0.4	n/a	0.08
	TLEV	0.125	3.4	0.4	15	n/a
	LEV	0.075	3.4	0.2	15	n/a
	ULEV	0.040	1.7	0.2	8	n/a
100,000	Tier 1	0.310	4.2	0.6	n/a	n/a
	Tier 1 diesel option	0.310	4.2	1.0	n/a	n/a
	TLEV	0.156	4.2	0.6	18	0.08
	LEV	0.090	4.2	0.3	18	0.08
	ULEV	0.055	2.1	0.3	11	0.04

---

[80] California Air Resources Board: Notice Of Public Hearing To Consider The Adoption Of Amendments To The Low-Emission Vehicle Regulations, November 15, 2001, http://www.arb.ca.gov/msprog/levprog/test_proc.htm.

[81] California LEV Program, see website http://www.arb.ca.gov/msprog/levprog/levprog.htm. See also The California Low-Emission Vehicle Regulations, http://www.arb.ca.gov/msprog/levprog/test_proc.htm)

[82] See California Air Resources Board, "California Exhaust Emissions Standards and Test Procedures for 2001 and Subsequent Model Passenger Cars, Light-duty Trucks, and Medium-Duty Vehicles," Proposed Amendments (Sept. 28, 2001).

Table 2-11. LEV II Exhaust Emission Standards for New MY2001-MY2003 Passenger Cars and Light-duty Trucks (8,500 lbs. GVW or less)						
Durability of Vehicle	Vehicle Emission Category	NMOG (g/mi)	Carbon Monoxide (g/mi)	NOx (g/mi)	Formaldehyde (mg/mi)	Particulates fr. diesel vehicles (g/mi)
50,000	LEV	0.075	3.4	0.05	15	n/a
	LEV Option 1	0.075	3.4	0.07	15	n/a
	ULEV	0.040	1.7	0.05	8	n/a
120,000	LEV	0.090	4.2	0.07	18	0.01
	LEV Option 1	0.090	4.2	0.10	18	0.01
	ULEV	0.055	2.1	0.07	11	0.01
	SULEV	0.010	1.0	0.02	4	0.01
150,000 (optional)	LEV	0.090	4.2	0.07	18	0.01
	LEV Option 1	0.090	4.2	0.10	18	0.01
	ULEV	0.055	2.1	0.07	11	0.01
	SULEV	0.010	1.0	0.02	4	0.01
	LEV	0.090	4.2	0.3	18	0.08
	ULEV	0.055	2.1	0.3	11	0.04

Zero Emission Vehicle (ZEV) Mandate

California's LEV regulations seek to push motor vehicle technology to its limits by introducing the Zero Emission Vehicle requirement, which began with LEV I and was amended for LEV II. Known as the "ZEV Mandate," this requires that a specific minimum percentage of passenger cars and the lightest light-duty trucks marketed in California by large or intermediate volume manufacturers be ZEVs.[83] With the adoption of the newer LEV II regulations, ZEVs considered in the program now include several classes of vehicles, including:

- Pure ZEVs (ZEVs)—vehicles with no tailpipe emissions whatsoever;
- Partial ZEVs (PZEVs)—vehicles that qualify for a partial ZEV allowance of at least 0.2 (before an additional "early introduction phase-in multiplier" or "high-efficiency multiplier" are applied to the allowance); and
- Advanced Technology PZEVs (AT PZEVs)—any PZEV with an allowance greater than 0.2.[84]

Pure ZEVs must produce zero exhaust emissions of any criteria or precursor pollutant under any and all possible operational modes and conditions. AT PZEVs include compressed natural gas, HEVs, and methanol fuel cell vehicles. In order to qualify as a PZEV, the AT PZEVs would also have to meet the SULEV tailpipe emissions standard,

---

[83] California Air Resources Board, "Notice Of Public Hearing To Consider The Adoption Of Amendments To The Low-Emission Vehicle Regulations," (November 15, 2001).

[84] California Air Resources Board, "California Exhaust Emission Standards and Test Procedures for 2003 and Subsequent Model Zero-Emission Vehicles, and 2001 and Subsequent Model Hybrid Electric Vehicles, in the Passenger Car, Light-duty Truck, and Medium-Duty Vehicle Classes," (Amended: April 12, 2002), pages A,B-1 to A,B-2. (hereinafter *California Exhaust Emission Standards and Test Procedures*). Qualified PZEVs meet SULEV, evaporative emissions, and on-board diagnostic standards, and offer an extended warranty of 15 years or 150,000 miles, whichever occurs first. See Id., page C-4.

achieve zero evaporative emissions, and include a 150,000-mile warranty for emission control equipment.[85]

The total required volume of a manufacturer's production and delivery for sale of Passenger Cars (PCs) and Light-duty Trucks 1 (LTD1s) is based on the average from the previous three-year period. The original LEV I regulations required that specific percentages of all PCs and LDT1s, MY1998 and later, be certified as ZEVs.

General Motors, DaimlerChrysler, and affiliated car dealerships in the state have brought a lawsuit against the CARB that may significantly change the course of the ZEV program, as discussed in the following sections. The ZEV program as currently adopted would required that 10 percent of all 2003 and subsequent model years be ZEVs, with a gradual *increase* in the minimum required percentage of ZEVs in sales fleets up to 16 percent by 2018.[86] As of Summer 2002, these most recent June 1, 2001 amendments are still pending, but are expected to be adopted without significant additional changes.[87] Pursuant to the lawsuit, a U.S. District Court Judge has issues a preliminary injunction that would delay implementation of the rules by two years, as of the writing of this report.

The most recent ZEV amendments require large and intermediate volume manufacturers to meet different percentage of sales requirements for pure ZEVs, PZEVs, and AT PZEVs.[88] Major automakers (those selling 35,000 or more passenger cars and light-duty trucks annually in California) could meet the 10 percent requirement for ZEVs sold in the State by selling 20% of their ZEV vehicles as pure ZEVs, 60% as PZEVs, and 20% as AT PZEVs. Intermediate automakers (those selling 4,501 to 35,000 passenger cars and light-duty trucks annually in California) could meet their entire ZEV requirement with PZEV credits, and manufacturers selling fewer than 4,500 vehicles annually would not have to meet any ZEV requirement.[89] Table 2-12 summarizes these requirements. (Note, small and independent low volume manufacturers are exempt from the ZEV requirements but can acquire credits for the sale of ZEVs or PZEVs.)

Table 2-12   Summary of ZEV Requirements under LEV II[90]

Applicable Manufacturer	Model Year	Percentage of Sales Required for Compliance
Large Volume Manufacturers	2003-2008	20% of sales as ZEVs (or ZEV credits)
		at least 20% of sales in additional ZEVs or AT ZEVs (or credits for such vehicles)
		remaining percentage (up to 60%) of sales as PZEVs (or PZEV credits)
Intermediate Volume Manufacturers	2003 and afterwards	up to 100% PZEV allowance vehicles (or credits)
Small Volume and Independent Low Volume Manufacturers	No requirements, but can acquire credits for sale of ZEVs or PZEVs	

Prior to the current lawsuit, the newly proposed regulations would also push back the start date for several requirements, such as the number of PZEV vehicles required in the

---

[85] California Air Resources Board, Zero Emission Vehicle Program Changes, "Fact Sheet" (December 10, 2001) http://www.arb.ca.gov/msprog/zevprog/factsheet/zevchanges.pdf. Note, the current Toyota Prius and Honda Insight HEV models do not yet meet all of the requirements needed to earn either PZEV or AT-PZEV credits. Id.

[86] The California Low-Emission Vehicle Regulations, http://www.arb.ca.gov/msprog/levprog/test_proc.htm.

[87] Telephone interview with Tom Evashenk, Staff, CARB (March 5, 2002).

[88] *California Exhaust Emission Standards and Test Procedures*, page C-2.

[89] SB 1782 (1998), see http://www.fleets.doe.gov/fleet_tool.cgi?$$,benefits.

[90] *California Exhaust Emission Standards and Test Procedures*, page C-2.

early years. PZEVs can now be phased in at 25 percent of the previously required level in 2003, and 50 percent, 75 percent, and 100 percent of the previous level in 2004, 2005, and 2006, respectively. Beginning in 2007, automobile manufacturers must also include heavier SUVs, pickup trucks, and vans in the sales figures used to calculate each automaker's ZEV requirement. In other words, in order to sell more SUVs and other heavier vehicles, each automaker must also sell more ZEVs.[91]

To date, nine NGVs have been developed that meet the California LEV and ZEV standards, as shown in Table 2-13. The Civic GX, for example, emits just one-tenth of the emissions permitted at the Federal ULEV standard (arguably emitting air that is cleaner than the ambient air in some cities). Today, the Civic GX remains the only vehicle certified in California as an "AT-PZEV".[92]

Table 2-13 Examples of Light-duty NGVs Meeting California Emission Standards as of 2001[93]		
Make and Model	**California LEV II**	**Fuel Displacement**
Passenger Cars		
Honda Civic GX CNG	AT-PZEV, SULEV	1.7 L
Ford Crown Victoria CNG	ULEV	4.6 L
Trucks, Vans, and SUVs		
Chrysler/Dodge Ram Van 2500	SULEV	5.2 L
Chrysler/Dodge Ram Van 3500	SULEV	5.2 L
Chrysler/Dodge Ram Wagon 2500	SULEV	5.2 L
Chrysler/Dodge Ram Wagon 3500	SULEV	5.2 L
Ford E-250	SULEV	5.4 L
Ford E-350	SULEV	5.4 L
Ford E-150	SULEV	5.4 L

Like the Federal Alternative Fuel Vehicle program under EPAct, the California program includes a range of credits that provide incentives for the development of ZEV vehicles with improved range and refueling capacity. Automakers will receive four times the normal number of credits for each ZEV introduced in 2001-2002, and 1.25 times the normal number of credits for each ZEV introduced between 2003 and 2005. The provisions also reduce the minimum number of extra credits available for ZEV models with extended ranges of 50 or more miles to 100 or more miles, and provide 10 credits for ZEVs with ranges of 275 or more miles. Extra credits are also awarded for vehicles that can refuel or charge in less than 10 minutes for a 60-mile range. ZEVs that remain on the road in California for more than three years also receive additional credits.[94]

[91] California Air Resources Board, Zero Emission Vehicle Program Changes "Fact Sheet," (December 12, 2001) http://www.arb.ca.gov/msprog/zevprog/factsheet/zevchanges.pdf.

[92] Natural Gas Vehicle Coalition, "Natural Gas Vehicles: The Environmental Solution Now," http://www.ngvc.org/ngv/ngvc.nsf/bytitle/environmentalbenefits.html. See also *Honda Civic GX CNG*, http://www.ngv.org/ngv/ngvorg01.nsf/bytitle/HondaCivicGX.htm; California Natural Gas Vehicle Coalition, *Honda Civic GX*, http://www.cngvc.org/ngv/CNGVC.nsf/bytitle/hondacivicgx.htm; and Honda Motor Corporation, http://www.hondacars.com/models/natural_gas_civic/fleet_info/?3.

[93] California Air Resources Board, "Buyer's Guide to Cleaner Cars," http://www.arb.ca.gov/msprog/ccbg/ccbg.htm.

[94] California Air Resources Board, Zero Emission Vehicle Program Changes, "Fact Sheet" (December 10, 2001), http://www.arb.ca.gov/msprog/zevprog/factsheet/zevchanges.pdf.

On January 3, 2002, the General Motors, DaimlerChrysler, and seven California car dealerships filed a lawsuit against CARB claiming that the ZEV program establishes fuel economy standards that are preempted by Federal law. In 2001, CARB approved amendments to the ZEV rules that included options for meeting ZEV requirements based on a vehicle's ability to reduce emissions as well as the vehicle's fuel economy. On June 11, 2002 a Federal District Judge in the Eastern District of California issues a preliminary injunction enjoining CARB from implementing and enforcing the ZEV program requirements for 2003 and 2004 model year vehicles. In its ruling, the court found that the ZEV rules did relate to fuel economy and were preempted by the Federal Energy Policy and Conservation Act of 1975 that establishes corporate average fuel economy (CAFE) standards, [95]

The effect of this ruling is that the ZEV program will be suspended for up to two years, pending an appeal by CARB that could change the ruling. As an alternative, CARB may also consider revising its rules to remove the fuel economy provisions, but it is unclear whether and when it would do so and what affect that would have on the injunction, if any. [96]

2.2.2 Adoption of California LEV II Standards in Northeastern States

As discussed above, California is the only State with the ability to adopt motor vehicle emissions standards that exceed those of the CAA. [97] However, under Section 177 of the CAA other States are permitted to adopt any regulations to address motor vehicle emissions that are enacted and adopted by California, so long as the regulations are no more stringent than California's standards and the regulations come into effect no sooner than two years after the applicable model year.

In the early 1990s, New York, Massachusetts, Maine, and Vermont adopted the California LEV standards. With the exception of Maine, which has repealed its California-based ZEV regulations, [98] each of those states has adopted the 10 percent ZEV sales mandate commencing in model year 2005, two years after the California start year of 2003. In 2000 and 2001, respectively, New York and Massachusetts took the further steps of adopting California's LEV II regulations, as amended. [99] Vermont has yet to adopt the most recently amended LEV II regulations, but is expected to do so. Beginning in model year 2005, New York also will require the LEV II program for medium-duty

---

[95] State Bar of California, Environmental Section, *Environmental Law News Update: Recent Judicial, Legislative, and Regulatory Developments* (July 2002), http://www.calbar.org/enviro/update/up0207.htm. See *Central Valley Chrysler-Plymouth, Inc., et al. v. California Air Resources Board and Michael P. Kenny,* Case No. F-02-05017 (E.D. Cal. 2002).
In a previous lawsuit brought directly by General Motors, the principal argument made by the plaintiffs is that, despite a multi-year effort, there has been "minimal market appeal of electric vehicles based on cost, range, and infrastructure issues." The complaint in that lawsuit alleged that the ZEV sales quota for battery-powered vehicles violated the California Environmental Quality Act. General Motors, *Latest News: GM Seeks Review of ZEV Mandate* (February 23, 2001). See *General Motors Corp. v. California Air Resources Board,* Cal. Super. Ct., No. C 01-00741, 2/23/01.
[96] Telephone interview with Tom Evashenk, Staff, CARB (August 26, 2002).
[97] 42 U.S.C. 4709(b).
[98] See State of Maine Department of Environmental Protection, Rule Chapter 127, *New Motor Vehicle Emission Standard,* Basis Statement for Amendments of December 21, 2000.
[99] In 1993, Maryland and New Jersey also adopted the California LEV program, provided that surrounding States also adopt the California standards. EVAA, State Laws and Regulations Impacting Electric Vehicles (January 2002), http://www.evaa.org.

vehicles, including larger pick-up trucks and SUVs weighing between 8,500 and 14,000 pounds.[100]

To date, New York and Massachusetts have adopted regulations that would provide automobile manufacturers greater flexibility in complying with the ZEV mandate. Manufacturers can choose to comply with either the California ZEV mandate beginning in model year 2005, or can opt into what is called the northeast states' ZEV Alternative Compliance Plan (ACP) in model year 2004, as explained in Table 2-14 below. In either case, manufacturers will be required to implement the full California ZEV mandate in model year 2007.[101]

Table 2-14	Summary of Alternative Compliance Plan for ZEVs in New York and Massachusetts[102]	
Model Year	Type of Vehicle	Percentage Requirements
2004	PZEVs	10% of all vehicle sales
2005	PZEVs	9% of all vehicle sales
	AT PZEVs or pure ZEVs	1% of all vehicle sales
2006	PZEVs	7% of all vehicle sales
	AT PZEVs	2% of all vehicle sales
	pure ZEVs	1% of all vehicle sales
2007	PZEVs	6% of all vehicle sales
	AT PZEVs	2% of all vehicle sales
	pure ZEVs	2% of all vehicle sales

2.2.3   Other State Programs

Other states have instituted a wide range of policy measures and programs designed to promote the use of AFVs. Such programs include tax incentives such as credits or deductions for AFVs and clean fuel equipment, exemptions from parking fees, special access to high occupancy vehicle (HOV) lanes, and other measures. Such measures include:

- individual or business tax incentives, including tax credits or deductions, for the purchase of AFVs and LEVs (AZ, GA, KS, LA, ME, MD, NY, OK, OR, UT, VA);
- individual or business tax incentives, including tax credits or deductions, for the construction of AFV and LEV fuel delivery systems (AZ, LA, RI, VA);
- tax incentives, including tax credits or deductions, for manufacturers of AFVs and LEVs (AK, MI);
- tax credits for each job created in manufacturing clean fuel vehicles or converting vehicles to operate on clean fuels (VA);

---

[100] Governor: Regulation to Reduce Harmful Vehicle Emissions, Alternative to Promote Clean Vehicle Technology, Improve Air Quality (January 4, 2002), http://www.state.ny.us/governor/press/year02/jan4_02.htm; See also New York Adopts New California Emission Standards, EarthVision Environmental News, November 29, 2000, http://www.climateark.org/articles/2000/4th/nyadnewc.htm.

[101] See Background Document and Technical Support For: Public Hearings on the Amendments to the State Implementation Plan for Ozone; and Hearing and Findings under the Massachusetts Low Emission Vehicle Statute - 310 CMR 7.40: The Massachusetts Low Emission Vehicle Program (February 2002), http://www.state.ma.us/dep/bwp/daqc/daqcpubs.htm.

[102] Governor: Regulation to Reduce Harmful Vehicle Emissions, Alternative to Promote Clean Vehicle Technology, Improve Air Quality (January 4, 2002), http://www.state.ny.us/governor/press/year02/jan4_02.htm.

- exemption of state and/or local sales tax for the purchase of AFVs or AFV conversion equipment (AZ, NH, PA);
- adjustments to fuel taxes to reflect use of AFVs (HI);
- grants to businesses, individuals, local governments, and non-profit organizations towards the purchase of AFVs or AFV fleets (AZ, CA, PA);
- regulations to facilitate the commercialization of AFVs (NH);
- requirements for state and municipal fleets to acquire AFVs and LEVs, to convert fleets to AFVs, to meet specific clean fuel standards, or to develop AFV infrastructure (DC, LA, MA, MI, MO, NV, NH, NM, NY, OK);
- exemption for certain AFVs or LEVs from emissions inspections and other motor vehicle registration fees and requirements (AZ);
- regulations addressing clean fuel vehicle identification labels or decals (CA);
- special requirements for public utilities to adopt and/or promote LEVs (CA); and
- research programs for the study of AFV technologies (SC, TN).

California Regulation of GHG Emissions from Motor Vehicles

On July 11, 2002, the California Legislature passed landmark legislation to propose adopting the first GHG emission regulations on motor vehicles in the United States. Signed into law on July 22, 2002 by the Governor of California, AB 1493 could significantly enhance the objectives of the State's LEV and ZEV program. The law requires the CARB to adopt regulations for carbon dioxide emissions from passenger cars, light trucks, and SUVs by January 1, 2005. The bill directs the CARB to adopt regulations "that achieve the maximum feasible reduction of GHGs emitted by passenger vehicles and light-duty trucks and any other vehicles" in the state. [103] The law would take effect January 1, 2006 and would apply to vehicles manufactured in the 2009 model year and after. One interesting condition in the final legislation is to require CARB to develop regulations that specifically do not: (1) impose additional fees or taxes on motor vehicles, fuel, or miles traveled; (2) ban the sale of any vehicle category in the state; (3) require reductions in vehicle weight; (4) limit speed limits; or (5) limit vehicle miles traveled. AB 1493 would also require the California Climate Action Registry to develop procedures by July 1, 2003, in consultation with CARB, for the reporting and registering of vehicular GHG reductions to the Registry. (The California Registry is described in greater detail in Section 2.3.1, below.) As stipulated in the Clean Air Act, once AB 1493 is signed into law, other states would be able to follow California in adopting equally stringent regulation of carbon dioxide emissions from automobiles.

California Clean Fuel Availability Requirements

To help promote the use of natural gas and other alternative fuels (including methanol, ethanol, and propane) in California, the CARB adopted additional rules requiring owners or operators of fuel stations to install fueling facilities at their stations. Under the regulations, for example, once vehicle manufacturers produce 20,000 dedicated NGVs, this would "trigger" the requirement for installation of fueling facilities. [104] The provisions cease to apply to each designated clean fuel once the number of retail clean fuel outlets offering the designated clean fuel represent at least 10 percent of all retail gasoline outlets in the state. [105]

---

[103] California, AB 1493 (as amended, May 31, 2001).
[104] Final Regulation Order, Amendments to the Regulations for the California Clean Fuels Program, see 13 CCR §§2300-2317.
[105] 13 CCR §2318.

The State of New York has taken several aggressive steps to promote the use of AFVs. On June 10, 2001, New York's Governor Pataki signed Executive Order No. 111 in an effort to exceed Federal AFV acquisition requirements under EPAct. Executive Order No. 111 requires all state government entities to met new acquisition requirements, regardless of the size of the fleets or where they are located. (Specialty, policy, and emergency vehicles are exempted.) By 2005, at least 50 percent of all new light-duty vehicles acquired by each fleet must be AFVs. After 2005, annual acquisition requirements must increase by 10 percent each year until 2010, when 100 percent of all new acquisitions will be AFVs.[106]

Maryland and Washington, DC AFV Purchase Incentives

Under the Maryland Department of Transportation Advanced Technology Vehicle Program, administered by the Metropolitan Washington Council of Governments, public and private fleets in Maryland and Washington DC may receive up to $4,000 per each dedicated AFV. Maryland also offers a tax credit of up to $2,000 for purchases of AFVs as a percentage of the Federal tax credit (discussed above). Vehicles weighing up to 5,000 pounds receive a credit of up to $800.[107]

2.3   Voluntary Programs and Support Activities Promoting NGVs

2.3.1   Voluntary Greenhouse Gas Registries and Reporting Programs

Over the last decade, various initiatives to register, document and promote voluntary GHG emission reduction measures have been introduced in the U.S., many of which parallel both voluntary and mandatory programs in other countries for complying with the Kyoto Protocol. The goal of these programs is to encourage public and private entities to participate in GHG reduction activities and to test procedures for GHG emissions accounting. Each program affords individual project developers with the opportunity to register and document activities that help reduce GHG emissions and to possibly use the registered emission reductions for participation in a future emissions trading regime.

The different programs range in scope and project type, and do not all include activities related to transportation. Two leading programs—DOE's 1605(b) Program and the California Climate Action Registry—are described below. Various other State GHG emissions registries have also been proposed, as well as an alternate Federal registry under the new EPA Climate Leaders Program.[108] Appendix 3 lists several new and proposed State initiatives to register GHG emission reductions, many of which encourage the development of GHG reduction measures that include the increased use of NGVs.

U.S. Department of Energy's 1605(b) Voluntary Reporting of Greenhouse Gases Program

Managed by the Energy Information Administration at DOE, the 1605(b) Voluntary Reporting of Greenhouse Gases Program (created under Section 1605(b) of EPAct)

---

[106] New York State Energy Research and Development Authority, *Executive Order No. 111 "Clean and Green" State Buildings and Vehicles Guidelines* (December 2001), p.12.
[107] Metropolitan Washington Council of Governments, *The Clean Alternative*, http://www.mwcog.org/trans/cleantaxi.html.
[108] Id.

affords any company, organization, or individual with the opportunity to establish a public record of their GHG emissions, emission reductions, and/or sequestration achievements in a central and public database. The program first began accepting reports on GHG reduction activities during calendar year 1995 and was among the world's first registries set up to track voluntary GHG reduction activities.

Like other registries, 1605(b) lays the foundation for maintaining information about individual projects, and standardizing GHG emissions accounting methodologies, which in turn makes possible the creation of a market wherein GHG emission reduction credits can be traded. Reporters generally participate in the program to gain recognition for environmental stewardship, demonstrate support for voluntary approaches to achieving environmental policy goals, support information exchange, and inform the general public about GHG reduction activities. If the participant has the emissions reductions certified by an independent third party entity, and the reductions meet the standards of a given emissions trading regime, then the participant may trade the certified credits within that regime and reap the financial benefits associated with the sale of those credits at market price. One example of such a regime, although still starting up, is the Chicago Climate Exchange, described in Section 2.3.2, below.

Data from the most recent 1605(b) reporting cycle, covering activities through 2000, were released by EIA in February 2002 and include considerable information on real-world transportation projects. Of the 72 transportation projects reported to the program, 15 were NGV projects involving direct emissions reductions of roughly 8,574 metric tons of carbon dioxide equivalent (CO_2E). Appendix 2 presents summary information on these projects, including the entities that undertook and reported the project, the name, scope and general description of each project, and the methods used to estimate the achieved GHG emission reductions. The data reported to the program is publicly available on DOE's website and may be useful for educational and project replication purposes.[109]

In recent years the 1605(b) Program has faced growing concern that it has not provided adequate standards that show true GHG emission reductions in participating organizations, and that the program would be unable to support GHG emissions trading. Starting with President Bush's February 2002 announcement, the 1605(b) Program is poised to be significantly strengthened and redesigned to serve as what may become the leading national system for tracking emissions and emission reduction activities, and perhaps establishing credits. Thus, the standards and methodologies that it establishes may become the default national standard that other registries and reporting programs, such as California's described below, may be integrated with it.

California Climate Action Registry

Largely in response the criticisms that the current 1605(b) Program lacked the necessary rigor that would be required in a GHG emissions trading program, in September 2001 the California Senate passed Senate Bill 1771 to establish the California Climate Action Registry—a non-profit organization providing a central and standardized system for reporting annual GHG emissions reductions. In return for voluntary registration of GHG emissions, the Registry promises to use its best efforts to ensure that participating organizations receive appropriate consideration under any future international, federal, or state regulatory regimes relating to GHG emissions.[110] Given the steps, described in Section 2.2.3, that California is taking to address vehicular GHG emissions in

---

[109] See http://www.eia.doe.gov/oiaf/1605/frntvrgg.html. For more information, contact the 1605(b) Program Communications Center at: 1-800-803-5182 or visit
http://www.eia.doe.gov/oiaf/1605/frntvrgg.html.

[110] California Energy Commission, Global Climate Change & California,
http://www.energy.ca.gov/global_climate_change/index.html.

transportation, the Registry may gain increased prominence for transportation related activities. As discussed above, AB 1493 directs the California Climate Action Registry to develop procedures for reporting and registering vehicular GHG reductions to the Registry.

In contrast to the 1605(b) program, entities participating in the California Registry have to report on *all* their emissions and emission reductions. At this point in time, the Registry does not accept reports that only include project-specific activities. Companies that wish to report on their transportation-related activities therefore also have to complete an inventory of company-wide emissions before submitting a report to the Registry.[111]

2.3.2   Emerging Markets for Trading GHG Credits

Another development, that is likely to have a significant impact on the development of GHG-related transportation projects and the increased market penetration of NGVs is the emergence of a new market for trading in GHG emission reduction credits. Though few governments have imposed binding restrictions on GHG emissions, many companies have already begun exploring the benefits and challenges of GHG trading.[112] The demand for AFV credits has been steadily increasing, driven in part by the anticipation of one or more regulatory regimes, and by the desire to earn a reputation as an environmentally conscious entity. As a result, a small but growing market for the sale and transfer of credits based on GHG reduction activities has evolved over the past few years. As this market continues to grow, opportunities for selling and trading credits derived from GHG reduction activities in the transportation sector will also increase. Potential GHG reduction opportunities that could be generated and sold for credit on the GHG market include projects promoting the use of cleaner vehicle options.

Trading activities have evolved in concert with a series of programs designed to help stakeholders gain experience and explore ways to address the climate change issue cost-effectively. These programs and initiatives have focused largely on project-level actions, and have included: US Initiative on Joint Implementation (USIJI); Ontario's multi-stakeholder Pilot Emissions Reduction program (PERT); the Dutch government's Emission Reduction Unit Procurement Tender (ERUPT); the World Bank's Prototype Carbon Fund (PCF); and the Oregon Climate Trust. While some programs, such as the Oregon Climate Trust, are not specifically trade-oriented, they are building experience in understanding the value of GHG offsets and the cost of implementing GHG mitigation activities, and such offset projects can potentially be sold on the GHG trading market. For example, the State of Oregon requires new power generators built in the state to avoid, sequester, or displace a portion of their previously unregulated carbon dioxide emissions, and have the option of providing funding to the Oregon Climate Trust (as a small percentage of total capital costs) to fund offset projects.[113]

Since there is no central recording entity for tracking GHG emissions trades, the actual size of the market is not fully known. However, as of 2001 it is estimated that approximately 65 inter-company transactions have occurred since 1996, involving roughly 50 to 70 million metric tons of CO_2E emissions reductions, although because many trades have occurred privately this number may be conservative.[114] The price of

---

[111] See California Climate Action Registry, http://www.climateregistry.org.

[112] Only the United Kingdom and Denmark have established formal emissions trading programs as a component of domestic climate change policies. The European Union is now preparing rules for an EU-wide GHG trading program, which is expected to enter into operation in 2005, and Japan recently announced it is developing a trading program.

[113] See Oregon Climate Trust website, as http://www.climatetrust.org.

[114] Richard Rosenzweig, Matthew Varilek, Ben Feldman, Radha Kuppalli, and Josef Jansen, *The Emerging International Greenhouse Gas Market* (Pew Center on Global Climate Change, March 2002).

public trades has ranged between \$0.60 and \$3.50 per metric ton of CO_2E. Most of these trades have been between buyers and sellers in Europe and North America, and the majority of trades have been verified by third-party, independent entities.

The most popular trading activities have included fugitive gas capture from landfills, fuel switching, energy efficiency, and co-generation.[115] None of the trades have involved reductions from transportation activities, highlighting the lack of experience with generating project-based GHG emission reductions in the transportation sector. However, as it is fairly straightforward to monitor and demonstrate ownership for this type of reductions it is likely that the types of activities traded will expand to include emission reductions from transport projects.

Chicago Climate Exchange

The Chicago Climate Exchange is emerging as one of the key U.S. organizations for helping to generate a viable trading market for GHG emissions reduction credits. In June 2001, 33 companies with assets in the Midwestern United States (including the Ford Motor Company) announced the formation of the Chicago Climate Exchange (CCX). Led by Environmental Financial Products and the Kellogg Graduate School of Management at Northwestern University, the group will explore the potential for a regional GHG trading exchange in order to achieve a specified level of emission reductions. The CCX has proposed that participating companies voluntarily commit to emissions reductions and trading in six GHGs.[116] Participants would commit to reducing their GHG emissions by two percent below 1999 levels by 2002 and reduce them 1 percent annually thereafter. Credits would be given for domestic and international emissions offsets projects after particular monitoring, verification, tracking and reporting requirements have been fulfilled. Potential emission reduction activities that could receive credit under the Chicago Climate Exchange include projects that reduce emissions from the transportation sector. Sample project types suggested by the CCX include fuel switching and vehicle efficiency improvement projects.

The CCX hopes to have the exchange up and running by the third quarter of 2002 for participants in seven states, including Illinois, Indiana, Iowa, Michigan, Minnesota, Ohio, and Wisconsin. In 2003, the CCX aims to have commitments and trading among participants in the entire United States, Mexico, and Canada, and to expand the exchange to include international participants in 2004.[117]

2.3.3 Other Programs Promoting NGVs

DOE's National Energy Technology Laboratory (NETL) Strategic Center for Natural Gas (SCNG) and Office of Fuels and Energy Efficiency

Noting that "Our Energy Information Administration tells us that natural gas will be the "fuel of choice" for the next 10 or 20 years, perhaps loner," former Secretary of Energy Bill Richardson established the Strategic Center for Natural Gas within NETL in December of 1999. The Secretary also reaffirmed that "We are counting on it (the

---

[115] Review and Analysis of the Emerging International Greenhouse Gas Market. Executive Summary of a confidential report prepared for the World Bank Prototype Carbon Fund. Natsource, 2001.

[116] The six gases covered by the CCX are carbon dioxide (CO_2), methane (CH_4), nitrous oxide (N_2O), hydrofluorocarbons (HFCs), perfluorocarbons (PFCs), and sulphur hexafluoride (SF_6).

[117] For more information on the Chicago Climate Exchange contact info@chicagoclimateX.com. Chicago Climate Exchange, 111 W. Jackson, 14th Floor, Chicago, Illinois 60604 USA. Phone: 1 (312) 554-3350, Fax: 1 (312) 554-3373, website: http://www.chicagoclimatex.com.

Center) to meet many of our energy goals <u>and</u> many of our environmental goals." He charged the National Energy Technology Laboratory (NETL) with creating a single center within the Department of Energy (DOE) to look out for the future of natural gas "from borehole to burnertip." With its 60-year history in gas production, processing and utilization, NETL was uniquely qualified to serve as the focus for DOE's natural gas research, development, and demonstration activities and was asked by the Secretary to "look to the big picture and devise the bold ideas that allow the full potential of natural gas to be achieved."

NETL also contributes to its commitment to promoting natural gas through the Office of Fuels and Energy Efficiency. This office operates programs in natural gas processing, transportation fuels and chemicals, advanced fuel research, and energy conservation programs. These programs develop economically sound technologies to provide cleaner transportation fuels, lower cost chemical manufacturing processes, and environmentally responsible use of fossil fuels. They also promote energy efficiency and sustainable development. The Office of Fuels and Energy Efficiency is implementing these goals by providing research and technical assistance to industry, government-industry partnerships, and other DOE offices.

A fundamental mission of DOE is to secure increased, reliable, and low-cost energy supplies while protecting the environment. Increased utilization of natural gas is a key element in achieving this goal. As a result, NETL works with industry, other DOE offices, and the National Economic Council of the White House to develop and implement a strategic plan for natural gas that promotes expanded gas use. The integrated plan removes redundancies and fills gaps in the current suite of DOE activities, and it ensures that all of DOE's work makes sense in the context of the entire natural gas system.

NETL focuses research into exploration and production, transmission and distribution, markets, and end-use technologies as well as the policy and regulatory framework of the nation's natural gas systems. While transportation is one of the smaller applications of natural gas use on the United States, NETL is committed to promoting and advancing NGV use and technology. One specific area of NETL focus is in gas-to-liquid (GTL) conversion research. The goal of that effort is to develop and demonstrate advanced technologies and processes for economic conversion of methane to liquids that can be used as fuels or chemical feedstock. This will increase the supply of liquid transportation fuels, thus reducing the demand for crude oil-derived transportation fuels.

DOE Clean Cities Program

Sponsored by DOE, the Clean Cities Program is designed to promote public-private partnerships to deploy AFVs and their supporting infrastructure. By encouraging AFV use, the Clean Cities program helps to achieve energy security and environmental quality goals on local, national, and international levels. Two principal goals of the program are to deploy one million AFVs operating exclusively on alternative fuels by 2010, and to promote one billion gasoline gallon equivalents of clean fuels used in AFVs by 2010.[118]

The Clean Cities program takes a voluntary approach to AFV development, working with coalitions of local stakeholders to help develop local strategies and initiatives to integrate AFVs into the local transportation sector. Participating cities in the program include:

- 77 Clean Cities coalitions in 41 states;

---

[118] See Clean Cities website, at http://www.ccities.doe.gov.

- 3 border programs with the cities of El Paso, Texas and Juarez, Mexico; Detroit, Michigan and Toronto, Canada; and Grand Forks, North Dakota and Winnipeg, Canada; and

- International programs in Chile, Brazil, Central America and Caribbean, India, Mexico, Peru, and the Philippines.

The DOE Clean Cities International Program began as a result of the Hemispheric Energy Symposium held in October 1995 in follow-up to the December 1994 Summit of the Americas to promote energy cooperation and sustainable development.[119]

U.S. Department of Transportation Programs

In May 1999, the U.S. Department of Transportation (DOT) announced that it was forming the Center for Global Climate Change and Environment to conduct scientific research on emerging technologies and alternative fuels to deal with carbon dioxide emissions from transportation sources. To address transportation issues related to climate change and global warming, officials from DOT said that the research center would focus on new technologies to achieve higher fuel efficiency, tax credits for fuel-efficient cars, changes in travel behavior, and transportation planning as part of community development. During the opening session, former Transportation Secretary Rodney Slater noted that transportation accounts for 26 percent of U.S. GHG emissions and that the new center would work closely with the Environmental Protection Agency and the Department of Energy to promote the development of low-emitting transportation technologies.

2.4   International Climate Change Programs

Although International agreements on the control of GHG emissions are not legally binding on domestic activities at this time, the international framework provides insight into the direction domestic legislation may take. This section begins with a description of international legislative developments that could have an influence on the number of climate change mitigation projects using NGV technologies.

2.4.1   International Framework for Promoting GHG Emission Reduction Projects

United Nations Framework Convention on Climate Change

International efforts to limit the release of GHGs[120] gained momentum at the United Nations Conference on Environment and Development (UNCED) held in Rio de Janeiro, Brazil in June 1992. This conference proved to be a turning point in the effort to reduce GHGs as well as the first international commitment to take specific actions to limit national emissions—the UNFCCC. Under the UNFCCC, industrialized countries voluntarily agreed to reduce their GHG emissions.[121] The U.S. Government ratified the UNFCCC on October 15, 1992 and is therefore considered a "Party" to the Convention. The Parties to the Convention meet every year at the ministerial level (Conference of the

---

[119] For more information, see http://www.ccities.doe.gov. Project developers may also contact the Clean Cities Hotline at 1-800-CCITIES for additional information.

[120] The most common anthropogenic (human-caused) greenhouse gases are CO_2, CH_4, N_2O, PFCs, HFCs and SF_6. There are other gases that trap heat in the earth's atmosphere, however the six gases (and classes of gases) mentioned above are those currently covered by international treaty.

[121] United Nations Framework Convention on Climate Change (UNFCCC), http://www.unfccc.int/resource/conv/conv_002.html.

Parties (COP)) and more often at the technical level to oversee and guide the implementation of the UNFCCC. From these annual COP meetings, and other meetings held by the subsidiary bodies to the UNFCCC, come most of the guiding international framework under which nations endeavor to limit GHG emissions.[122]

Several initiatives have been proposed under the UNFCCC to promote and credit project-based GHG reduction activities. In 1995, the Parties to the Convention established the Activities Implemented Jointly (AIJ) Pilot Phase, under which a framework was developed for implementing emission reduction projects jointly between two or more countries. This concept is generally known as joint implementation (JI). Moreover, the Kyoto Protocol to the UNFCCC was produced in 1997 during COP-3, which took place in Kyoto, Japan.[123] The Protocol language establishes legally binding emission reduction targets for industrialized countries. The Kyoto Protocol establishes two project-based mechanisms that could encourage the development of NGV-related projects in exchange for certified emission reduction units: the Clean Development Mechanism (CDM), involving projects between actors in an industrialized and a developing country; and Joint Implementation (JI), involving projects between actors in industrialized countries.[124]

United States Domestic Climate Change Policy

In 2001, the United States announced its intention to fully withdraw from the Kyoto Protocol process and would not ratify the treaty. With his announcement, President Bush offered that the U.S. would develop an alternative approach to reducing domestic GHG emissions. The key components of this domestic policy are still under development, but generally include the following:

- a commitment to reduce GHG emissions intensity—the ratio of GHG emissions to economic output—by 18 percent over ten years;
- improvements to the U.S. national GHG emissions registry (reporting) program, known as the Voluntary Reporting of GHGs "1605(b)" Program (established under Section 1605(b) of EPAct), now implemented by the Energy Information Administration (EIA) in DOE;
- protection and provision of transferable credits for GHG emission reductions under a future climate change regime; and
- a commitment of financial and technical resources for the continued research of climate change and innovative new technologies to reduce GHG emissions.[125]

It is important to note that, in addition to these recent domestic policy activities, increased international activity to implement the Kyoto Protocol could be a potentially important driver for increased development and implementation of NGVs in overseas markets.

---

[122] For more information on the COP meetings, see http://unfccc.int. See also International Institute for Sustainable Development, http://www.iisd.ca/climate.

[123] The Kyoto Protocol to the United Nations Framework Convention on Climate Change, http://www.unfccc.de/resource/docs/convkp/kpeng.html.

[124] The Protocol must be ratified by 55 Parties to the Convention, representing at least 55 percent of Annex I parties' 1990 carbon dioxide emissions, before it can enter into force. (Annex I parties are generally considered industrialized countries or countries with economies in transition.) As of the date of this publication, over 85 Parties have signed and ratified the Kyoto Protocol, representing 37 percent of global emissions, and several countries that have not yet ratified the treaty have expressed the likelihood of doing so. The Protocol will enter into force on the ninetieth day after the date on which no less than 55 Parties to the Convention, incorporating Annex I Parties which account for at least 55 percent of the total carbon dioxide emissions for 1990 from that group, have deposited their instruments of ratification, acceptance, approval or accession. See http://unfccc.int/resource/kpstats.pdf.

[125] National Oceanic and Atmospheric Administration, President Announces Clear Skies & Global Climate Change Initiatives (Silver Spring, Maryland, February 14, 2002), available at: http://www.whitehouse.gov/news/releases/2002/02/20020214-5.html.

This, in turn, could have meaningful effects on the relative availability and cost of NGV products that can subsequently be used in the U.S.—particularly in States pursuing California LEV II motor vehicle standards.

Joint Efforts under the UNFCCC

In addition to taking on voluntary reduction targets, the Parties to the UNFCCC agreed to develop national programs to slow the release of harmful emissions and to take climate change into account in such matters as agriculture, energy, natural resources, and activities involving coastal areas. The Parties also agreed to share technology internationally and to cooperate in other ways to reduce GHG emissions, especially in the energy, transport, industry, agriculture, forestry, and waste management sectors. Together, these sectors produce nearly all of the GHG emissions that can be attributed to human activities. However, the Convention does not establish legally binding emission reduction requirements for the signatories.

As a result of the potential GHG emission benefits associated with switching from gasoline to natural gas fueled vehicles, the promotion of NGV projects would support many of the major goals set forth in the convention. For instance, the development of individual NGV projects and the adoption of policies to promote the use of NGVs would greatly enhance national efforts to limit emissions of GHGs. Industrialized countries could also fulfill their commitment to share technology and cooperate with other nations by facilitating the transfer of NGV technologies to developing countries, for example through participation in AIJ projects.

2.4.2   Activities Implemented Jointly (AIJ) Pilot Phase

The UNFCCC introduced the concept of JI, which refers to arrangements through which an entity in one country partially meets its domestic commitment to reduce GHG levels by financing and supporting the development of a project in another country. To test the concept of JI, the AIJ Pilot Phase was established at the first Conference of the Parties to the UNFCCC (COP-1), held in Berlin in 1995. Projects initiated during this phase were called "activities implemented jointly" to distinguish them from the full-fledged JI projects the Convention was considering for future implementation. The goal of the AIJ Pilot Phase was to provide developing nations with advanced technologies and financial investment while allowing industrialized nations to fulfill part of their reduction commitment at the lowest cost. Because of the temporary pilot status of this program, it was decided that project developers would not receive credit or other monetary incentives for projects developed and approved as part of this initiative.

The Parties adopted three basic criteria for the pilot phase of AIJ:

1. The activity must be officially approved as an AIJ project by both countries involved;

2. The activity must result in real, measurable and long-term reductions in net GHG emissions that would not have occurred in the absence of such an activity; and

3. The activity should be financed outside current Official Development Assistance (ODA) funds.

Although the AIJ Pilot Phase provides a unique opportunity for the development and recognition of NGV-related GHG emission reduction projects, few such projects have actually been implemented. The two most common types of AIJ projects are land use (including forest conservation, forestry, and sustainable forest management) and energy (primarily stationary combustion and fuel switching projects). Of the 156 AIJ projects currently approved by the designated national authorities for AIJ, one is a transportation project, which involved fuel switching from diesel to natural gas.[126]

A single transportation project has been approved under the AIJ Pilot Phase to test and advance NGV technologies in Hungary. This project, which is called the RABA/Ikarus Compressed Natural Gas Engine Project, is being carried out between project developers in the Netherlands and Hungary. The goal of the project is to replace about a thousand public transport diesel buses with new CNG buses, and to promote technology transfer to Hungary to assist two vehicle manufacturers, RABA and Ikarus, in building and delivering new CNG vehicles to the Hungarian market. The project is also expected to build market potential for the Dutch companies involved, as well as to strengthen the economic position of the Hungarian companies receiving the technology transfer. The initial cost estimate of GHG emission reductions resulting from the project ranges between $100 and $250 per ton of CO_2 equivalent reduced. Approximately 39 percent of the initial funding for this project was to come from the Dutch Government, another 39 percent from the Hungarian Government, and the remaining 22 percent from other Hungarian sources.[127] At the time of AIJ approval, the project was expected to achieve 7,400 tons of CO_2 reductions per year and to continue to achieve reductions for over 20 years.[128]

At least one other transportation-related AIJ project has been considered for development. Also a natural gas vehicle project, this AIJ project would be conducted between project developers in the United States and Chile and would switch 100 buses or taxis from diesel or gasoline to natural gas. The limited number of transportation-related AIJ projects is by no means a reflection of the transportation sector's share of total GHG emissions (which is significant). Recognizing the need to exploit more fully the opportunities offered by low-emitting transportation technologies, the countries promoting participation in the AIJ Pilot Phase are eager to help facilitate development of projects in the transportation sector. In this connection, NGV projects are often cited as relevant mitigation activities due to the many local energy and environmental side benefits of such projects.

Since the initiation of the AIJ Pilot Phase, a number of countries, including Costa Rica, Japan, Norway, Poland, Sweden, Switzerland, and the U.S., have established national offices to facilitate and evaluate AIJ projects. In addition, several other countries have identified a designated focal point within their governments to oversee project development and approval.[129]

---

[126] For more information, see the UNFCCC's Activities Implemented Jointly website, http://www.unfccc.de/program/aij/aijproj.html.

[127] Estimated emission reductions were derived using data on the numbers and types of buses initially in the Hungarian fleet; emissions data for vehicle and engine types were derived from a standardization emission test, fuel consumption data, and an estimated average of 65,000 kilometers driven per bus per year. The full project description is posted on the UNFCCC website at http://www.unfccc.int/program/aij/aijact/hunnld01.html

[128] See http://unfccc.int/program/coop/aij/aijproj.html.

[129] See http://unfccc.int/program/coop/aij/aij_np.html

3 GHG Emissions And Natural Gas Vehicles

3.1 Introduction

In developing NGV GHG emission reduction projects, project developers should have a thorough understanding of the procedures for quantifying the resulting GHG emissions reductions. Quantification is one of the first and necessary steps in developing a project that may later qualify for crediting in a market-based program or regime (i.e. where the credits may be monetized and/or traded). The following subsections provide an overview of the issues related to estimating and documenting the potential GHG emission reductions achieved by replacing conventional gasoline or diesel powered vehicles with NGVs.

This chapter is divided into the following subsections:

- The types of domestic and international transportation-related GHG reduction projects that have been undertaken and for which data has been reported;
- Types and sources of GHG emissions associated with NGVs;
- studies and models that may be helpful in estimating emission reductions from vehicle projects;
- Common rules and procedures for quantifying and documenting GHG emission reduction activities under project-based GHG mitigation programs; and
- Emerging framework for quantifying GHG emission reductions from transportation projects.

This chapter will be followed by a case study, presented in Chapter 4, in which the procedures for quantifying GHG emission reductions are illustrated. Thus, the principles described herein will form the basis for understanding the various issues involved in the actual accounting methodologies required for eventual crediting.

3.2 Projects Deploying NGV Technologies to Reduce GHG Emissions

There are five main types of activities that can be undertaken to reduce GHG emissions in the transportation sector. These include:

- *Changing vehicle fuel type:* sample activities include switching from gasoline or diesel to alternative fuels such as biodiesel, natural gas, electricity and hydrogen (via fuel cells or direct combustion);
- *Changing vehicle fuel efficiency:* changing engine and vehicle design such that the vehicle will travel further for the same amount of fuel or input energy;
- *Increasing vehicle occupancy rate:* This is another type of efficiency improvement where the number of person- or cargo-miles is increased for a given quantity of fuel or input energy. Activities may include car sharing, telematic systems for freight, or subsidized public transport.

- *Mode switching to less GHG-intensive transportation options:* increased public transportation, light rail systems, etc.; improved traffic management/infrastructure changes; and
- *Reducing transportation activity:* e.g. road pricing (congestion charges) and telecommuting;

Each option focuses on different ways to reduce emissions, ranging from behavioral changes to direct substitution of transport technologies. Hence, the procedures for estimating and accounting for emission reductions are different for each of the five activity types.

For the individual NGV project developer, the option of fuel switching from gasoline or diesel to natural gas by switching or upgrading to vehicles that run on natural gas is the most relevant, as it refers to activities that can be undertaken directly by the individual fleet manager. For example, by replacing a fleet of gasoline-powered taxis with NGVs, a fleet manager will reduce GHG emissions by taking advantage of the fact that natural gas has a lower carbon content than gasoline. Other transportation activities, such as increased use of public transit, improved traffic management, telecommuting, vehicle retirement, carpooling, or improved road or rail infrastructure, would mostly involve behavioral or regulatory changes that would likely be implemented by public authorities, automobile manufacturers, or private companies seeking to reduce the transport activities of their employees. While these measures may be equally or more effective in reducing GHG emissions, they are less relevant in the context of this paper.

While there is much domestic and international experience to speak of with regard to alternative fuel vehicle programs and site-specific projects, there is very little experience in developing and implementing these projects with the specific aim of reducing and accounting for GHG emissions.[130] This is an important distinction, because it illustrates where the industry "learning" needs to take place. In many cases, fleet operators and public planners are well aware of the various technology options and tools to provide incentives for AFVs, but they are likely unfamiliar with the rules and modalities of accounting for and documenting GHG emissions resulting from AFV projects in a transparent and accurate manner. This should come as no surprise, as there has been little impetus in the form of regulations or binding GHG reduction commitments imposed on companies or governments. However, as the concern for climate change and the likelihood of some form of future regulatory action grows, so does the need to accurately quantify and document GHG emissions reductions, especially if these reductions are to be used to meet future regulatory requirements, or are to be traded in a market based regime. Furthermore, given the disparate nature of the sources of GHG emissions relative to, for example, those in the power sector (millions of mobile, individual vehicles versus a handful of stationary plants), policymakers have focused on other sectors which would more readily lend themselves to GHG reduction projects and accounting. Thus, it is only recently that general, recognized methodologies for GHG accounting from transportation activities have been developed.

As a measure of the lack of experience at the international level, it should be noted that, of the 157 projects registered with the UNFCCC Secretariat as AIJ pilot projects, only one deals with the transportation sector (the RABA/IKARUS Compressed Natural Gas Engine Bus project – funded by Dutch investors and hosted in Hungary[131]). In the U.S., the number of voluntary actions to reduce GHG emissions in the transportation sector is also

---

[130] Most project-based GHG reduction activities target sectors such as electricity generation, industrial energy use, renewable energy development, or land use and forestry activities.

[131] The project involves the development and testing of a new CNG engine to be installed by the companies of RABA and Ikarus in new buses.[131] These buses will replace the purchase of 1,500 diesel buses. http://www.unfccc.int/program/aij/aijproj.html.

low. In 2002, there were 72 transportation related GHG emissions reduction projects reported to the DOE Voluntary Reporting of Greenhouse Gases Program—a small number compared to the 462 electricity generation, transmission, and distribution projects reported for the same year.[132] Nearly half (31) of these transportation projects involved AFVs, and 15 involved the use of NGVs. Finally, one transportation project has received carbon financing through the Climate Trust, a non-profit organization developed by the state of Oregon to help power plants meet state-specified CO_2 efficiency standards.[133] The project accepted by the Climate Trust involves internet-based carpool coordination in the Portland area and is expected to reduce 70,000 metric tons of CO_2 by eliminating 161 million vehicle miles traveled over 10 years.[134] In comparison, several hundreds of GHG reduction projects have been developed in the energy, industrial, and land-use sectors.

Concerned with the lack of transportation sector projects to mitigate GHG emissions, various governments and project-based GHG offset programs have been promoting their development. For example, DOE issued a grant in the fall of 2000 to the Washington D.C.-based Center for Sustainable Development in the Americas (CSDA) to create an AIJ project using natural gas vehicles in Santiago, Chile. Moreover, the World Bank's PCF is actively working with developing country representatives to identify transportation projects that may be suitable for carbon financing.[135]

3.3   GHG Emissions Associated with NGVs

CO_2 is by far the largest emission source related to transportation because it is the natural result of combustion of carbon-based fuels such as gasoline and diesel (both of which have high carbon contents relative to natural gas). Methane, or CH_4, is also an important transportation-related GHG, but not because it is a product of combustion. Instead, CH_4 is a component of many fuels (the primary component of natural gas – usually comprising around 95%, depending on the blend), and its delivery to the atmosphere takes place via leakages in the infrastructure and fuel delivery systems, including leakages from the vehicle itself. Furthermore, since CH_4 has 23 times the Global Warming Potential[136] of CO_2 over a 100-year time frame, smaller volumes of emissions will have magnified climate change consequences. N_2O is also a greenhouse gas associated with transportation emissions, but accounts for a very small amount of overall emissions, despite having a GWP or 210.

As is illustrated in Figure 3-1, developed by the Argonne National Laboratory (ANL) using the Greenhouse Gases, Regulated Emissions, and Energy Use in Transportation (GREET) Model (also described in Section 3-4 below), light-weight natural gas vehicles realize significant reductions (17%) in life-cycle CO_2 emissions relative to equivalent light-

---

[132] Energy Information Administration. http://www.eia.doe.gov/oiaf/1605/frntvrgg.html.

[133] "Funding Internet-Based Carpool Matching with CO2 Offsets" presentation International Council for Local Environmental Initiatives (ICLEI), Seattle Workshop, February 8, 2002. Presented by Kris Nelson, the Climate Trust: http://www.climatetrust.org

[134] Based on an estimated 2% per year increase in car and vanpools. CarpoolMatchNW.org

[135] The PCF is a World Bank fund that seeks to "buy" high quality, low cost GHG reduction credits from international project developers. High quality refers to the degree to which the resulting credits are environmentally additional (see section 3.5.1 below), transparent and verifiable, and how well the project achieves the environmental and sustainable development goals of the host country.

[136] GWP of a GHG is the degree to which that gas will enhance the overall effect of global warming. It is a function of the gas' radiative forcing potential (or how well the gas transmits visible radiation and traps infrared radiation). GWP is expressed in relative terms, with CO_2 as the base, for a given period of time. The concept of GWP allows for the comparison of emissions of different GHGs, such as CH_4 and CO_2, using a common unit: tons of CO_2 equivalent (tCO2E). CH_4 has a GWP of 23, meaning that one ton of methane emitted will act as if 23 tons of CO_2 had been emitted.

weight gasoline vehicles with similar engine efficiencies due to the lower carbon content of natural gas. However, the overall reduction in GHGs is discounted some (down to 11%) due to the methane emissions from the leakage associated with natural gas.

For medium- and heavy-duty vehicles, the GHG reductions are less clear, since these vehicles use diesel engines, which are typically more efficient than gasoline engines. This is discussed further in the paragraphs that follow.

Figure 3-1.   Changes in Fuel-Cycle GHG Emissions Relative to Gasoline Vehicles Fueled with Clean Gasoline[137]

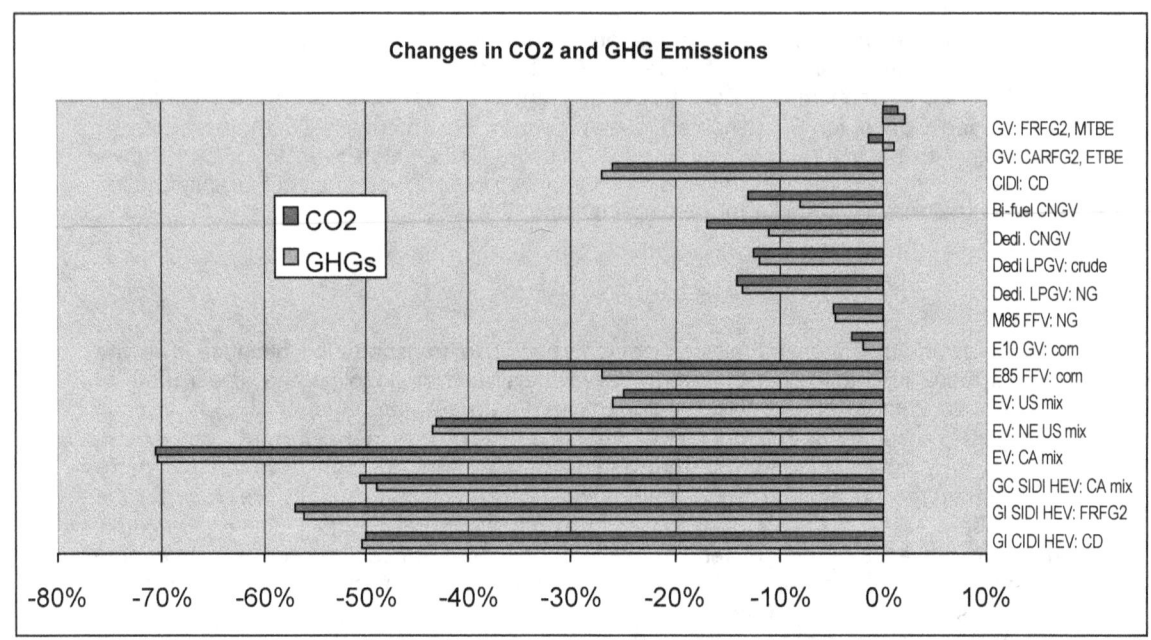

3.3.1   Carbon Dioxide Emissions

Although there are demonstrable CO_2 reductions from NGVs, these emissions benefits vary depending on the type of fuel and vehicle model being replaced. In general, *light-duty NGVs* have a significant CO_2 advantage relative to conventionally fueled vehicles (gasoline and diesel). For similar engine combustion efficiencies in light-duty vehicles, natural gas typically has a 20 to 40 percent tailpipe CO_2 emissions advantage versus conventional fuels.[138] This advantage stems from the lower carbon-to-hydrogen ratio

---

[137] Argonne National Laboratory. GREET 1.5—Transportation Fuel-Cycle Model. Volume 1: Methodology, Development, Use and Results. August 1999. GI=grid independent; CIDI=compression ignition, direct injection; FRFG2=Federal Phase 2 reformulated gasoline; SIDI=spark ignition, direct injection; E85=mixture of 85 % ethanol and 15% gasoline by volume; FFV=fuel flexible vehicle; E10=mixture of 10 % ethanol and 90% gasoline by volume; GV=gasoline vehicle; M85=mixture of 85 % methanol and 15% gasoline by volume; NG=natural gas; LPGV=liquefied petroleum gas vehicle; dedi=dedicated; CNGV=compressed natural gas vehicle; CD=conventional diesel; CARFG2=California Phase 2 reformulated gasoline; ETBE=ethyl tertiary butyl ether; MTBE=methyl tertiary butyl ether.

[138] James McCarthy and Sean Turner, "Natural Gas Vehicles and Greenhouse Gas Emissions." Presentation for the NETL-sponsored training session, *Developing International Greenhouse Gas Emission Reduction Projects Using Clean Cities Technologies*. San Diego, California, May 10, 2000.

characterizing natural gas (see Figure 3-2), which makes the CO_2 production from combustion of CNG and LNG relatively low compared to gasoline. However, as mentioned previously, this figure is discounted when all emissions are accounted for, due to the leakage of methane in the natural gas delivery infrastructure and the vehicle itself, such that the overall emissions benefits for a light-weight vehicle will be on the order of 17%. Emissions factors for each gas at each stage is listed in Table 3-1 below.

When natural gas vehicles are compared to *medium- and heavy-duty* diesel-fueled vehicles, which is the most commonly used fuel type for larger-sized vehicles, the CO_2 benefits are not as significant, despite the fact that the carbon content of diesel is much higher than that of natural gas (see Figure 3-2). This is because diesel vehicles typically have more efficient engines and thus use less fuel per mile traveled than equivalently-sized NGVs. As a result, the reduced fuel consumption of diesel vehicles per mile traveled offsets some of the CO_2 benefits derived from the lower carbon content of natural gas. The exact difference in CO_2 emissions between the two fuel types will vary

Figure 3-2. Typical Carbon Input per Unit Energy

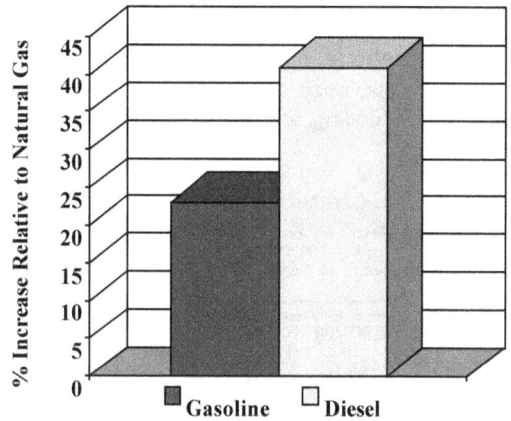

Source: GREET model. Argonne National Laboratory, August 1999.

depending on engine type and vehicle efficiency of the medium- and heavy-duty vehicles, which makes it difficult to make generalizations about the resulting GHG benefits. To estimate the net CO_2 emissions difference of medium- and heavy duty vehicles, it would be necessary to obtain information about the specific combustion efficiencies or fuel consumption of both the diesel and natural gas vehicles under consideration.

3.3.2 Methane Emissions

With a methane content of more than 90 percent, the amount of methane emitted from the use of natural gas vehicles is much higher than the level emitted from conventional gasoline vehicle. Emission factors used in the GREET model, which attempt to quantify the amount of methane emitted per mile traveled based on aggregate leakage data from all stages in the fuel cycle, are listed in Table 3-1 below. As the table shows, for light-duty vehicles, in the vehicle operation stage methane emissions from NGVs are 10 times higher than those in gasoline vehicles, and in the feedstock stage, NGV methane emissions are twice as high. The GREET model does not contain similar data for heavy-duty vehicles.

In spite of the significant contribution of methane to NGV-related GHG emissions, studies of transportation-related emissions often only refer to tailpipe emissions, which include CO_2, but not CH_4. This exclusion is primarily due to the higher degree of uncertainty associated with non-tailpipe emissions, which is amplified when dealing with medium- and heavy-duty vehicles, as less research has been done to quantify these emissions.

3.3.3 Life Cycle Analysis and GHG Emissions

Vehicle related GHG emissions arise from several stages of the fuel cycle. The GREET model separates vehicle emission sources into three categories, including:

- **Feedstock-related** – the production of the raw materials (e.g. crude oil) used to make the useful fuels (e.g. gasoline or diesel). This includes feedstock/resource recovery (production), and feedstock/resource transportation and storage.

- **Fuel-related** – conversion from a feedstock to a useful fuel, which includes refining, and the associated transportation, storage, and distribution; and

- **Vehicle operation** – includes the final preparation for delivery to the vehicle tank, such as the compression or liquefaction in the case of natural gas, vehicle refueling, and vehicle operation (including tailpipe emissions)

Figure 3-3. Contribution of Each Stage of the Fuel Cycle to Total Fuel-Cycle Energy Consumption and Emissions[139]

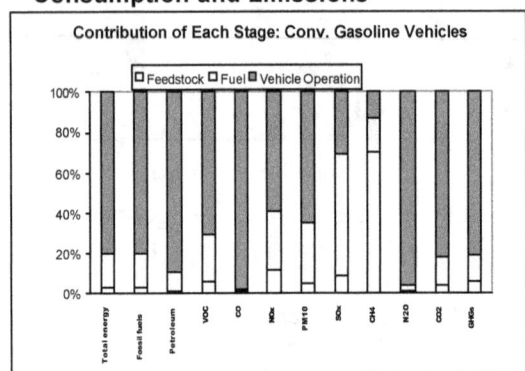

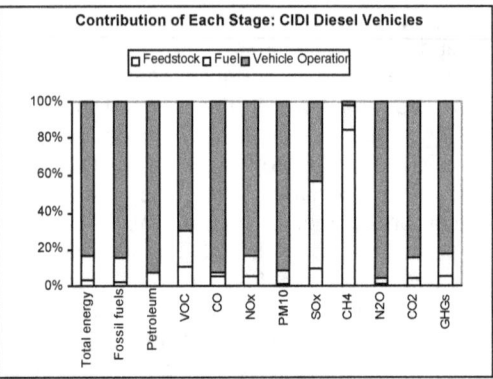

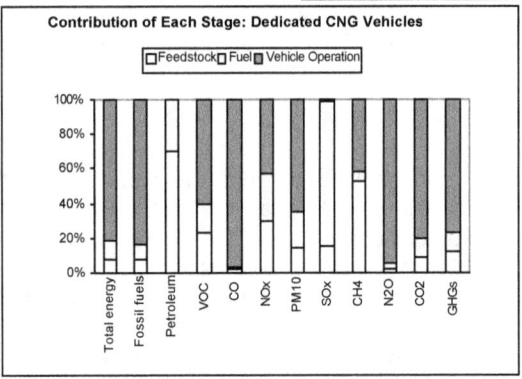

Figure 3-3 is an illustration of the relative share of emissions and energy use from the three stages of the fuel cycle, for each combination of fuels and vehicles. These figures are created from results of the GREET model and separate the fuel-cycle into the three stages listed above. The figures illustrate that to gain a comprehensive understanding of vehicle emissions, the full fuel cycle from "well to wheel" has to be considered. Table 3-1 lists the GHG emission factors on a per mile and per BTU basis, which led to Figure 3-1 and Figure 3-3.

---

[139] Michael Q. Wang, *GREET 1.5 – Transportation Fuel-Cycle Model: Volume 1, Methodology, Development, Use and Results.* Argonne National Laboratory, August 1999.

Table 3-1. Emissions Factors for Light-Weight Vehicles[140]

Emissions Factors per Distance Traveled (Mile)			
Conventional Clean Gasoline Vehicle			
	Feedstock	Fuel	Vehicle Operation	Total
CH4 (g/mi)	0.467	0.116	0.084	0.667
N20 (g/mi)	0	0.001	0.028	0.029
CO2 (g/mi)	18	82	390	490
GHG (gCO2E/mi)	28	85	401	514
Energy (BTU/mi)	192	1144	5156	6492

Emissions Factors per unit energy (MMBTU)			
Conventional GV: Clean Gasoline			
	Feedstock	Fuel	Vehicle Operation	Total
CH4 (kg/MMBTU)	2.43	0.10	0.02	0.10
N20 (kg/MMBTU)	0.00	0.00	0.01	0.00
CO2 (kg/MMBTU)	93.75	71.68	75.64	75.48
GHG (kgCO2E/MMBTU)	145.83	74.30	77.77	79.17

Compressed Natural Gas Vehicle			
	Feedstock	Fuel	Vehicle Operation	Total
CH4 (g/mi)	1.094	0.098	0.84	2.032
N20 (g/mi)	0.001	0	0.022	0.023
CO2 (g/mi)	38	41	330	409
GHG (gCO2E/mi)	61	43	355	459
Energy (BTU/mi)	533	625	5544	6702

CNGV			
	Feedstock	Fuel	Vehicle Operation	Total
CH4 (kg/MMBTU)	2.05	0.16	0.15	0.30
N20 (kg/MMBTU)	0.00	0.00	0.00	0.00
CO2 (kg/MMBTU)	71.29	65.60	59.52	61.03
GHG (kgCO2E/MMBTU)	114.45	68.80	64.03	68.49

It should be noted that a full life-cycle analysis adds considerable complexity to the emissions baseline estimation process, and may increase the potential for error and overall transaction costs of project development if the emissions factors are not developed based on country-specific circumstances. Life-cycle emissions and methodological procedures for collecting this data will likely vary from country to country due to differences in energy mix, fuel supply, and transportation characteristics. However, many countries in the developing world do not have the required data and institutional resources to undertake an adequate life-cycle analysis. This lack of data may limit the ability of project developers to accurately determine the emissions benefits of potential NGV projects.

One solution may be to exclude the full fuel cycle analysis from the baseline analysis and rely solely on tailpipe emissions data. As is illustrated in Figure 3-3, tailpipe emissions from vehicle operation comprise more than 75 percent of total GHGs from gasoline, diesel, and natural gas vehicles. Thus, a simplified baseline estimation process considering only tailpipe emissions will introduce errors no greater than 25 percent into the emission reduction estimates.[141] The effect of this potential error could be mitigated by discounting a similar percentage of the claimed emissions reductions, or by adding a predetermined grams/mile increment to the baseline calculation.[142]

3.4 Studies and Measurements of GHG Emission Benefits of NGVs

As GHGs are only regulated in a few countries, a limited number of studies and publicly available resources are available to offer assistance in estimating GHG emissions from vehicles. The following summaries provide an overview of the major information sources on GHG emissions benefits from NGVs. As the studies indicate, GHG emissions vary

---

[140] Michael Q. Wang, *GREET 1.5 – Transportation Fuel-Cycle Model: Volume 2, Appendix B, Per-Mile Fuel-Cycle Energy Use and Emissions.* Argonne National Laboratory, August 1999.

[141] Michael Q. Wang, "Fuel-Cycle Analysis of Transportation Fuels: Development and Use of the GREET Model," and James McCarthy and Sean Turner, "Natural Gas Vehicles and Greenhouse Gas Emissions," presentations for the NETL-sponsored training session, *Developing International Greenhouse Gas Emission Reduction Projects Using Clean Cities Technologies.* San Diego, California, May 10, 2000.

[142] See "Well-to-Wheel Use and Greenhouse Gas Emissions of Advanced Fuel/Vehicle Systems – North America Analysis," General Motors Corp, Argonne National Laboratory, BP, ExxonMobil, and Shell. April, 2001. www.powertrain.se/pdf/63.pdf

depending on the technology and vehicle category in question, especially when non-tailpipe emissions, including methane, are included in the analysis.

In general, the studies indicate that light-duty LNG or CNG vehicles result in substantially fewer GHG emissions than similarly sized conventional gasoline vehicles. However, when GHG emissions of heavy-duty vehicles are compared, the GHG benefits are less obvious (as was discussed in section 3.3 above).

3.4.1 Greenhouse Gases, Regulated Emissions, and Energy Use in Transportation (GREET)

The GREET model was developed to calculate fuel-cycle energy use (Btu/mi) and emissions (g/mi) for various fuels. It calculates emissions of five criteria pollutants and three GHGs, as well as use of total energy, fossil energy, and petroleum. GHG emissions for vehicles are easily calculated using the Model. It was developed by the Argonne National Laboratory to make calculations of the GHG emissions of light-duty conventional vehicles and alternative fuel vehicles in the U.S. All the GHG emissions from vehicle use and upstream from fuel production, are included. Three GHGs (CO_2, N_2O and CH_4) are combined with their GWPs to calculate CO_2-E GHG emissions. GREET also evaluates criteria pollutant emissions, and compares fuel efficiency and emissions for CNGs relative to conventional gasoline vehicles. Users are able to change the default values to accommodate their specific situation. The GREET model is free of charge and can be downloaded from: **http://www.transportation.anl.gov/ttrdc/greet.**

It should be emphasized that the model is based on U.S. conditions and energy infrastructure. Users from other countries should be careful to adopt model inputs, which are relevant to country-specific conditions. These should include country-specific assumptions regarding fuel use and GHG emissions during the production, refining, and transportation of fuels and the national electricity mix used for electricity generation.

3.4.2 Canada's Transportation Climate Change Table

In May 1998, Canada's federal, provincial, and territorial Ministers of Transportation established the Transportation Climate Change Table as part of the national process to develop a climate change strategy.[143] The Table was comprised of transportation sector experts from a broad cross-section of business and industry, government, environmental groups and non-governmental organizations. It was mandated to identify specific measures to mitigate GHG emissions from Canada's transport sector.

The Transportation Climate Change Table submitted its Options Paper, "Transportation and Climate Change: Options for Action" to the Ministers of Transportation and the National Climate Change Secretariat in November 1999. The Options Paper assesses the costs, benefits and impacts of over 100 measures. The Transportation Table undertook 24 studies in support of the Options Paper. As part of this effort, several alternatives to gasoline, including NGVs, were compared for their potential for reductions in GHG emissions.[144] The comparison included vehicle efficiencies (miles traveled per BTU of fuel energy input), upstream emissions from fuel production, per-mile GHG emissions, fuel costs, and projected vehicle costs 10 and 20 years in the future.

---

[143] For more information on Canada's Transportation Climate Change Table visit the website at: http://www.tc.gc.ca/envaffairs/english/climatechange/ttable/ or email: TCCTable@tc.gc.ca.

[144] Alternative and Future Fuels and Energy Sources For Road Vehicles. Prepared for Canada's Transportation Issue Table, National Climate Change Process. Levelton Engineering Ltd. in association with (S & T)² Consulting Inc., BC Research Inc., Constable Associates Consulting Inc., Sierra Research.

http://www.tc.gc.ca/envaffairs/subgroups/vehicle_technology/study2/Final_report/Final_R eport.htm.

3.4.3 The Environmental Protection Agency (EPA) MOBILE6 Model

The emission rates of local air pollutants of AFVs and engines are readily available from EPA. The EPA's MOBILE6 model allows fleets to calculate the emissions reductions they can be expected to result from real-world operation of AFVs. MOBILE6 is a computer program that estimates hydrocarbon, carbon monoxide (CO), and nitrogen oxide (NOx) emission factors for gasoline and diesel fueled highway motor vehicles, as well as for AFVs such as natural gas and electric vehicles that may be used to replace them. MOBILE6 calculates emission factors for 28 individual vehicle types in low- and high-altitude regions of the United States. Emission factor estimates depend on various conditions, such as ambient temperatures, travel speeds, operating modes, fuel volatility, and mileage accrual rates. Many of the variables affecting vehicle emissions can be specified by the user, tailoring the calculations to specific types of fleets. MOBILE6 will estimate emission factors for any calendar year between 1952 and 2050, inclusive. Some states, such as California, have similar software which are specific to their unique climate and driving characteristics. Estimates of emissions reductions are often needed for AFV owners to apply for and receive grants from incentive programs.

EPA is undertaking an effort to develop the next generation of modeling tools for the estimation of emissions produced by on- and off-road mobile sources, including the New Generation Model. The new model will expand the scope of pollutants and GHGs covered and improve on the accuracy in calculating mobile source emissions, keeping pace with new analysis needs, new modeling approaches, and new data.

3.4.4 Additional References Addressing Emissions and Energy Use From Natural Gas or Alternative Fuel Vehicles

A number of additional studies provide useful information regarding potential GHG emission benefits from NGV and other vehicle project types. The most important studies include:

- "Well-to-Wheel Energy Use and Greenhouse Gas Emissions of Advanced Fuel/Vehicle Systems; North American Analysis," developed by the General Motors Corporation, the Argonne National Laboratory, BP, ExxonMobil, and Shell. April, 2001. Available at: www.powertrain.se/pdf/63.pdf
- "Life-cycle Emissions Analysis of Alternative Fuels for Heavy Vehicles," by Australia's Commonwealth Scientific and Industrial Research Organisation (CSIRO). March, 2000. Available at: http://www.greenhouse.gov.au/transport/pdfs/lifecycle.pdf
- "An Assessment of the Emissions Performance of Alternative and Conventional Fuels," from the UK Department of Transportation's Cleaner Vehicles Task Force. January, 2000. Available at: http://www.roads.dft.gov.uk/cvtf/
- "Saving Oil and Reducing CO_2 Emissions in Transport; Options and Strategies," from the International Energy Agency. 2001. Available at cost at: http://www.iea.org/public/studies/savingoil.htm.

3.5 Procedures for Estimating GHG Emissions Benefits from EV and HEV Projects

This section discusses some of the major issues related to the quantification of NGV-related GHG benefits, with the intent of instructing project developers on the steps and considerations in calculating GHG emission reductions. It is important to note at the outset that the quantification of emissions benefits is not synonymous with the certification of GHG reduction credits, although the two processes are certainly related. Quantification is one of the first and necessary steps in developing a project that will later qualify for crediting in a market-based program or regime (i.e. where the credits may be monetized and/or traded). As a general rule of thumb, when developing projects that are to eventually be certified and credited for registration and/or trading, and where the credits may be used to finance the project, project developers must first develop a plan that forecasts the emissions and emissions reductions that will take place as a result of the project. This plan would also include a methodology for data collection (in the case of a transportation project, this data would likely include fuel purchase records, odometer readings, vehicle maintenance records, etc.) so that the *actual* emission reductions can eventually be calculated. The awarding of credits will not take place until after the reductions have taken place (i.e. after the annual project data has been recorded, submitted and approved.)

The concept of a "GHG credit" implies the recognition by a specific GHG emission reduction regime or program based on a specific set of pre-determined criteria. The criteria will vary from one regime or program to another, and thus any discussion regarding the crediting process must either address the criteria of a specific program, or merely speak generally about the types of criteria. However, the elements of the quantification methodology, described in the paragraphs below and used in the case study in the following chapter are more or less common to the various nascent and evolving domestic and international programs that have been developed thus far to credit and register GHG reductions, and therefore are a good representation of the required steps for most programs. These common elements include: The *quantification* methodology is useful as an end in-and-of-itself as a calculation tool, but it is also a component within a larger GHG reduction crediting scheme. The ultimate specifics of the quantification methodology will depend on the program or regime in which the reductions are to be credited.

During the past decade, a series of project-based programs and initiatives have been introduced to gain experience and harness the power of markets in order to address the issue of climate change in a cost-effective manner. These programs, although governed by a unique set of rules, exhibit some common elements that constitute a *de facto* (though non-binding) set of minimum quality criteria that govern the creation of credible emission reductions Leading examples of these programs include: USIJI; the AIJ Pilot Phase, Canada's PERT in Ontario; Oregon's Climate Trust; ERUPT of the Dutch government; the World Bank PCF, and the Kyoto Protocol's project-based programs including Joint Implementation and the Clean Development Mechanism.

The following rules and issues are the common elements of project-based systems and provide a framework for project developers interested in developing GHG reduction projects.

3.5.1 GHG Emissions Baseline

The emissions baseline is an integral part of the GHG reduction project proposal because it is used to estimate emissions benefits of the project and will be used as the basis for awarding credits to the project. Many project-based programs measure emissions reductions by comparing the emissions performance of a credible "without project" baseline (i.e. the emissions that would have taken place if the project did not exist) against the "with project" emissions (or the emissions that actually do take place as a result of the project). The baseline—either static or dynamic—is used for comparison with emissions resulting from the project. The challenge of developing emissions baselines stems from the uncertainty of projecting what will happen in a given economy or specific market 10, 20, or 30 years in the future. Static baselines (Figure 3-4) rely on historical information to fix emissions at a set level, such as an entity or project's physical emissions in a given year. This same emissions level is then used every year throughout the life of the project as a reference to measure emissions reductions against.

Dynamic baselines (Figure 3-5) are emissions baselines that attempt to account for changes that may take place during the life of the project. As such, dynamic baselines are linked to particular variables and may be revised upward and downward depending on project and entity characteristics such as market penetration,

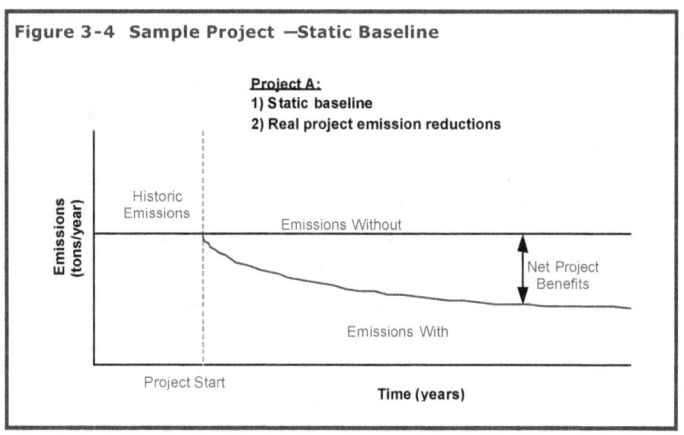

Figure 3-4 Sample Project —Static Baseline

Project A:
1) Static baseline
2) Real project emission reductions

growth rates, efficiency rates, and peer group benchmarks. For example, a law enacted sometime in the future mandating use of a given technology or fuel option will

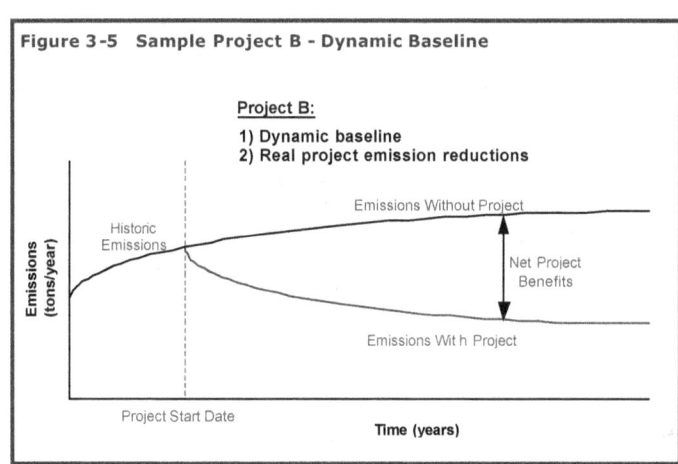

Figure 3-5 Sample Project B - Dynamic Baseline

Project B:
1) Dynamic baseline
2) Real project emission reductions

dramatically alter the use of that technology or fuel and associated emissions levels. In this case, an accurate baseline would either forecast such a law and its effects, or it would be revised to account for it.

Once the baseline has been determined, the estimate of emissions "with the project" can be developed. To determine project emissions, the same assumptions and time frames used for the "without project" baseline should be applied. Most project cases lead to *real* emission reductions. However, as illustrated in Figure 3-6, it is sometimes possible that actual emissions with the project will continue to rise above historical emissions. Such projects may still be able to obtain GHG reduction credits, as long as the reported project emissions performance continues to fall below the emissions associated with the baseline scenario.

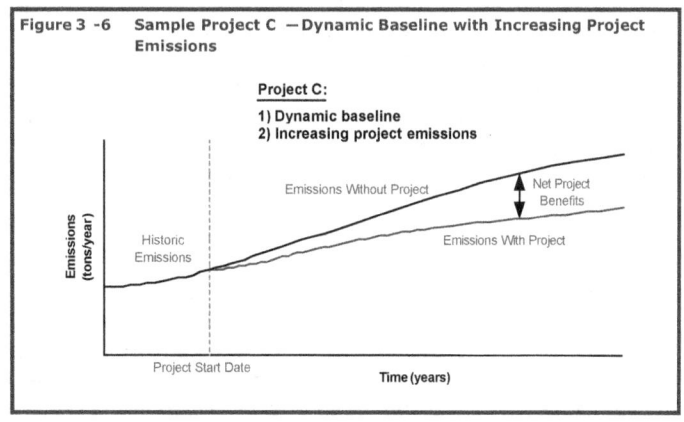

Figure 3 -6   Sample Project C —Dynamic Baseline with Increasing Project Emissions

Project C:
1) Dynamic baseline
2) Increasing project emissions

3.5.2 Environmental Additionality

The requirement of environmental additionality is linked closely to the process of developing the GHG emissions baseline. Environmental additionality is the requirement that emissions reductions achieved by a project must not have occurred in the absence of the project. That is, credits awarded to the project developers must stem from emissions reduction activities undertaken *in addition to* the business-as-usual scenario. To be credible, baselines should therefore take into account any laws, regulations, or technology improvements that may have a direct or indirect impact on GHG emissions.

In the case of NGVs, the question of additionality is more straightforward than is the case with other technology types in other sectors, due to the general lack of NGV market penetration and limited prospects for increased market penetration of NGVs in the near future. However, if the purchase of NGVs were mandated by an existing law or regulation, the baseline would need to account for it to be considered additional.

3.5.3 Leakage

Leakage is another common criterion that requires that the project developers provide evidence that the emissions reductions achieved at the project site do not lead to increases in emissions outside the boundaries of the project (i.e., emissions "leakage"), or that the calculation of claimed emissions reductions quantifies and accounts for leakage. Switching to electric vehicles is a good example of a project type with potential for leakage. If the boundary of the project is limited to an analysis of tailpipe emissions alone, the emissions will be reduced to zero, when in fact significant emissions may be produced at the power plant in the generation of the electricity for powering the electric vehicle. These power plant emissions would have "leaked" from the accounting system. Therefore, project developers should be careful to define the "boundary" of the project.

3.5.4 Ownership

Most programs require that the project developer, or those seeking claims to the resulting credits, has a legitimate claim to ownership of the reductions generated by the project and that other potential claimants are identified. Ownership can be demonstrated through documents certifying and dividing ownership clearly among all project participants. If necessary, supporting documents by local or national government authorities can be included to verify the validity of claimed ownership. The issue of ownership is an important consideration for transportation projects, especially in many countries where buses and taxis are owned by individual vehicle operators rather than one single fleet operator. When the ownership of a transportation project covering 200 vehicles is divided among a similar number of owners, contractual and other issues may become very complicated. One solution may be to form an association representing all the vehicle owners, which could then be listed as the owner of the project.

3.5.5 Monitoring and Verification

Another common requirement is that project developers develop a plan or procedure for outlining how emission reductions are to be monitored throughout the life of the project. The measured reductions must then be verified by an independent third party, who certifies that monitored reductions and/or the proposed method for calculating emissions performance can be or has been audited to provide a credible quantitative assessment of actual project performance. Both the monitoring and verification requirements involve guidelines for validating and verifying that no leakage will take place and that the GHG emissions baseline is estimated correctly (i.e. that the reductions meet the environmental additionality requirement).

3.6 Emerging Guidance for Estimating GHG Emissions From Transportation Projects

Although guidance has been available for estimating GHG emission benefits from energy, industrial, and waste projects for many years, procedures related to transportation projects are only just emerging. Even so, none directly address GHG emissions related to NGVs. However, useful indicators for how to calculate emissions from transportation projects can be derived from the World Resources Institute and World Business Council for Sustainable Development (WRI/WBCSD) Greenhouse Gas Protocol Initiative—a multi-stakeholder initiative between industry, government, and non-governmental organizations, to develop generally accepted accounting practices for measuring and reporting corporate GHG emissions.[145] The resulting standard and guidance are supplemented by a number of user-friendly GHG calculation tools (excel spreadsheets), which can be accessed on the GHG Protocol website (www.ghgprotocol.org).[146] Although the GHG Protocol focuses on corporate emissions, the proposed accounting standards and reporting instructions serve as an indicator of how project-specific emission reductions could be calculated.[147]

According to the protocol, there are two general methodologies for calculating emissions from vehicle projects: fuel-based and distance-based.[148] The preferred method is the "fuel-based" approach, which is based on previously aggregated fuel consumption data to determine emissions. Following this approach, fuel consumption is multiplied by the CO_2 emission factor for each fuel type in order to derive CO_2 emissions. The fuel emission factor is developed based on the fuel's heat content, the fraction of carbon in the fuel that is oxidized, and the carbon content coefficient. To calculate emissions the following equation should be used:

$$CO_2 \text{ Emissions} = \text{Fuel Used} \times \text{Heating Value} \times \text{Emission Factor}$$

In the case that project developers do not have access to site-specific information, default emission factors and heating values for different transportation fuels are listed in the

---

[145] *The Greenhouse Gas Protocol: A Corporate Accounting and Reporting Standard.* WRI/WBCSD. Washington, D.C. 2000. Under the GHG Protocol, corporate transportation emissions take the form of either direct or indirect emissions. Direct emissions refer to emissions that are associated with owned or controlled sources, such as company owned vehicle fleets and corporate aircraft. Indirect emissions refer to all other company-related emissions, including employee commuting, short-term vehicle rentals, and upstream/downstream transportation emissions. If companies purchase electricity for owned or operated EVs, the related emissions should be reported as indirect emissions and should use guidance developed in the 'Stationary Combustion Tool' for calculating emissions. For all other vehicles, including HEVs, companies should use the methodologies developed for calculating direct emissions from mobile sources.

[146] Only transportation-related CO_2 emission estimates are included in this tool. According to the GHG Protocol, accounting for N_2O and CH_4 emissions is optional at the discretion of the user. This is because N_2O and CH_4 emissions comprise a relatively small proportion of overall transportation emissions.

[147] The WRI/WBCSD is also in the process of developing a GHG project accounting model with the aim of developing general guidance for emission reduction and land use, land-use change and forestry projects. This module will include accounting procedures for transportation projects.

[148] "*Calculating CO₂ Emissions from Mobile Sources*" WRI/WBCSD GHG Protocol Initiative. www.GHGprotocol.org.

GHG Reporting Protocol guidance documents. Alternatively, emission factors such as those developed in the GREET model and presented above, can be used as well.

Fuel use data can be obtained from several different sources including fuel receipts, financial records on fuel expenditures, or direct measurements of fuel use. If specific information on fuel consumption is not available, information on vehicle activity data (i.e. distance traveled) and fuel economy factors (such as miles per gallon) can be used to calculate fuel consumption, using the following equation:

> Fuel Use = Distance x Fuel Economy Factor

The GHG Protocol also includes default fuel economy factors for different types of mobile sources and activity data.

The second methodology is the "distance-based" approach, and should only be used in the case where information on fuel use cannot be obtained. In the distance-based method, emissions are calculated by using distance-based emission factors to calculate emissions. Activity data could be expressed in terms of vehicle-kilometers (or miles) traveled, passenger-kilometers (or miles), and so on. This information is then multiplied by a default distance-based emission factor[149] according to the following equation:

> CO_2 Emissions = Distance Traveled x Distance-Based Emission Factor

Default distance-based emission factors are provided in the GHG Protocol guidance documents, and emission factors per distance driven and BTU of fuel energy, as developed in the GREET model, are available as well. Because this approach is based on default emissions factors it will be far more inaccurate and should only be used when the necessary data for the fuel-based approach is unavailable.

The WRI/WBCSD unfortunately does not include emissions factors and calculation procedures for examining CH_4 emissions from natural gas vehicles, and thus project developers may have to bolster the protocol with additional emission factors, such as those developed in the GREET model.

---

[149] A sample default distance-based emission factor could be 0.28 kg CO_2 per mile traveled for a small petrol car with no more than a 1.4 liter engine. "*Calculating CO2 Emissions from Mobile Sources*" WRI/WBCSD GHG Protocol Initiative. www.GHGprotocol.org.

4 Case Study on Quantifying GHG Emissions from NGVs

4.1 Introduction

In the previous chapter, some of the major issues related to the quantification of NGV-related GHG benefits were presented and discussed, with the intent of instructing project developers on the steps and considerations for calculating GHG emissions reductions. In this chapter, we take this discussion one step further and present a project case study in order to illustrate the process and familiarize the reader with the specific issues that should be considered during the quantification of GHG emission benefits.

The following subsections provide a brief summary of the project case study, outline the general criteria for developing a GHG reduction project based on current market-based proposals for GHG emissions control, develop the project based on these criteria, and estimate the emissions baseline and net project benefits. Three sample baseline scenarios are provided to illustrate how different project characteristics may influence the baseline estimate. The three baselines include:

(1) A static baseline focusing on tailpipe emissions;
(2) A dynamic emission baseline focusing on tailpipe emissions and changes to equipment over time; and
(3) A dynamic baseline, including full fuel cycle analysis and changes to equipment over time.

The case study is based on a hypothetical project that involves the deployment of 75 compressed natural gas taxis to replace 75 aging gasoline-fueled taxis.[150] The case study focuses on the process of developing an emissions baseline and estimating net GHG emission benefits of an individual project.

4.2 General Emissions Reductions Calculation Procedure

The first step is the estimation of what emissions would have been if the project had *not* been implemented. This step is also known as the emission baseline, the project reference case, or the business-as-usual scenario and should forecast emissions for the entire life of the project. Because the potential project emission benefits are derived by comparing actual or "with-project" emissions to the reference case, accuracy in the development of the reference case is very important. However, as discussed in Chapter 3, estimating future emissions is a difficult process. First, it is almost impossible to factor in everything that may or may not happen 10 to 20 years in the future, and several different (potentially equally valid) results are possible depending on which assumptions are used. GHG reduction programs and project developers planning to receive credit for their projects under a future market-based GHG reduction program must be careful to develop baseline criteria that would be stringent enough to be accepted under any program. Given the differences between the various initiatives to credit GHG reduction activities, developers should consult the preliminary guidelines of each of the proposed

---

[150] Julie Doherty and Jette Findsen, "Case Study: CNG Taxis, The Republic of Clean Cities." Presentation for the NETL-sponsored training session, *Developing International Greenhouse Gas Emission Reduction Projects Using Clean Cities Technologies.* San Diego, CA, May 10, 2000.

programs before developing a project, and be careful to detail all assumptions and emission sources when quantifying the potential emission benefits. The examples provided in the following case study are less comprehensive and should only be used as an indicator of the types of data and quantification procedures that could be required from the various GHG reduction programs.

The second quantification step involves *estimating* emissions with the project in place. This should include an estimation of all relevant "with-project" emissions for the entire life of the project. This estimation is done at the outset of the project during the project development and financing stages and, along with the baseline, completes the plan for estimating the emissions reductions that are likely to result from the project. As was mentioned earlier, for actual crediting to take place, estimates of emissions will need to be based on ex-post reported data (such as fuel purchase records or odometer readings). During the estimation of "with-project" or actual emissions, project developers should be careful to define the project boundary and detail all the assumptions and emission sources included in the estimate.

The third and final step in quantifying projected emissions reductions resulting from the project entails subtracting the actual or "with-project" emissions from the baseline or reference case emissions. The difference will represent the net emissions *reductions* of the project.

4.3 Emission Reduction Project for Taxis

This case study is based on a hypothetical project in a country called the Clean Cities Republic.[151] Although the Clean Cities Republic is a developing country, it does not represent any country or region in particular. The numbers used for this case study are fictional. The data provided for estimating the emissions baseline have been developed to illustrate how to quantify potential emission benefits. The data should not be used as an indicator of the specific emissions potential of an NGV project. Natural gas vehicle project developers should obtain their own GHG emission data for both the conventional vehicles to be replaced and the new alternative fuel vehicles to be introduced.

4.3.1 Republic of Clean Cities Background Information

The Republic of Clean Cities is a country with a population of 45 million people. Gross domestic product (GDP) is US$190 billion per year, and growing at a rate of 5 to 6 percent over the last 10 years. As a result of this economic expansion, the country is experiencing energy demand growth of 7 percent per year, with the transportation sector representing the fastest growing energy sector. Currently, transportation activities account for 32 percent of energy related CO_2 emissions, and this share is expected to grow significantly over the next few decades as the transportation sector continues to expand.

The project will be located in the capital of the Republic of Clean Cities, a city of 8 million people with population growth of 5 percent per year. On average, there are 7 people per motor vehicle, compared to 1.3 per vehicle in the U.S. The total number of vehicles on the road is growing by 7 percent annually. The capital is experiencing serious local air quality problems and is among the 20 most polluted cities in the world. The

---

[151] The hypothetical country example of the Republic of Clean Cities was first introduced at the 6th National Clean Cities Conference for illustrating a similar case study on estimating the GHG benefits of a natural gas vehicle project. Julie Doherty and Jette Findsen, "Case Study: CNG Taxis, The Republic of Clean Cities," Presentation for the NETL-sponsored training session, *Developing International Greenhouse Gas Emission Reduction Projects Using Clean Cities Technologies*, in San Diego, CA, May 10, 2000.

concentration of total suspended particulates (TSP) in the air is 8 times higher than the proposed World Health Organization (WHO) standards. The majority of the capital's pollution problems are caused by transportation-related emissions. To alleviate some of these environmental problems, the government has introduced tax incentives for switching to alternative fuel vehicles. In addition, a recently passed law mandates that all new cars should use unleaded gasoline. Currently, 40 percent of all gasoline sold in the country is leaded. The local government has also introduced a car use reduction plan to curb the rapid growth of new vehicles in the capital area. Finally, a new domestic regulation was put in place this year for reductions in vehicle tailpipe emissions of criteria pollutants.

The natural gas refueling infrastructure is still very limited in the capital as well as the rest of the Republic of Clean Cities. No CNG refueling stations have been introduced in the capital. However, a new pipeline was recently built for transporting natural gas to the capital. The recent construction of the pipeline ensures that leakage from the system is minimal. A portion of the natural gas supplied to the capital originates at an oil field where it was previously flared and/or vented into the atmosphere.

4.4 The Project Case Study

As part of the project, 75 dedicated CNG taxis (sedans) will be purchased to replace 75 aging gasoline taxis. To develop a supporting refueling infrastructure, one new CNG refueling station will be constructed at the site where these taxis are parked. Moreover, an extensive training course will be provided for the fleet mechanics. The lifetime of the project is estimated conservatively at 10 years. Each taxi is expected to drive an average of 80,000 miles per year. The total estimated GHG emission benefits of the project are expected to reach up to 11,687 metric tons of CO_2 equivalent.

The project participants include the Capital City Transportation Department, a local taxi fleet operator, and a U.S.-based NGV manufacturer. The CNG project has been approved by the Republic of Clean Cities' National Climate Change Office, which has been authorized by the Ministries of Foreign Affairs, Energy, and Environment to evaluate and certify AIJ and other international climate change projects. The National Climate Change Office, administered by the Ministry of Environment, has provided written documentation of project approval.

The project reduces CO_2 emissions by replacing gasoline with natural gas, and by reducing the need for oil recovery, gasoline refining, and fuel transportation (which produces more CO_2 emissions than the production and transportation of natural gas). The CO_2 savings offset the increased CH_4 emissions associated with natural gas recovery, pipeline leakage, natural gas compression, and fuel combustion. N_2O emissions remain mostly unchanged and will not be included in the emission baseline.

4.5 Project Additionality

Determining the additionality of the EV project is relatively straightforward. Although natural gas is cheaper than gasoline in the Republic of Clean Cities, the absence of a CNG refueling infrastructure has prevented, and will continue to prevent, vehicle operators from purchasing CNG vehicles. In particular, the high up-front cost of purchasing and installing a refueling station discourages the deployment of CNG vehicles in the capital. In addition, CNG vehicles have an incremental cost of between $3,000 and $5,000 per vehicle, adding another barrier to investment. This has led analysts to the conclusion that investment in CNG vehicles in the Republic will not take place without special incentives, legal requirements, or funding plans initiated by the government or some other funding source. As there are no such policies or funding initiatives currently under consideration by the local or national government authorities, it can be assumed that the deployment of CNG vehicles would not have happened without the prospect of obtaining future GHG mitigation-related credits. Therefore, it can be asserted that this NGV project is clearly additional.

If, on the other hand, NGVs had achieved significant market penetration among taxis, or a law were in place mandating that 15 percent of all public and private fleets must consist of low emission vehicles, such as natural gas vehicles, the issue of additionality would be less straightforward. In the latter case, the project developer would not be able to claim GHG emission reduction credits for NGVs purchased to meet the 15 percent requirement. Only vehicles purchased to exceed the mandated low emission requirements would receive credit. Hence, a fleet owner with 200 conventional gasoline taxis—who replaces 40 old conventional gasoline vehicles with 40 new NGVs—would only be able to obtain emission reduction credits for 10 of the new NGVs. The other 30 vehicles would go towards meeting the 15 percent mandate for zero emission vehicles. However, for the purposes of the following case studies it is assumed that no such laws have been put in place, and that no significant NGV market penetration exists.

4.6   Developing the Emissions Baseline

Since the introduction of the concept of cooperatively implemented GHG reduction projects, little experience has been gained regarding the development and evaluation of transportation-related GHG reduction projects. As mentioned earlier, only one transportation project has been approved under the UNFCCC's AIJ Pilot Phase. One project, however, does not provide enough precedent to be used for the development of standardized methodologies for analyzing transportation projects.

According to the criteria of USIJI, the emission baseline should include *major* emission sources and GHGs from the project.[152] For this type of project proposal it may be sufficient to include information about tailpipe emissions only, instead of completing an entire life cycle analysis. However, it is possible that any future market-based program would require a more stringent analysis of potential emission reductions, making inclusion of tailpipe CO_2 emissions only insufficient.

Because of the many unanswered questions related to the requirements of establishing an emission baseline, this study will provide three sample baseline scenarios that range from less detailed to very comprehensive in nature. The three baselines include:

1.   A static baseline focusing on tailpipe emissions;
2.   A dynamic emission baseline focusing on tailpipe emissions and changes to equipment over time; and
3.   A dynamic baseline including full fuel cycle analysis and changes to equipment over time.

The purpose of presenting these different baseline scenarios is two-fold. One is to advance the discussion on some of the issues that must be resolved in order to establish clear guidelines for the documentation and approval of transportation-related projects. The other purpose is to provide potential project developers with an idea of the issues that must be considered during the development of an emissions baseline for a transportation project. Project developers can then choose between or combine the

---

[152] *Resource Document on Project & Proposal Development under the U.S. Initiative on Joint Implementation (USIJI).* U.S. Initiative on Joint Implementation, Version 1, June 1997. Emphasis added.

different baseline scenarios depending on the purpose and requirements of the program to which the project participants will be applying for credit.

Factors that may determine the choice of baseline scenarios, include:

1. The transportation technology used for the project;
2. Availability of full fuel cycle and tailpipe emissions data;
3. Individual GHG program requirements;
4. The risk tolerance and level of accuracy desired by project developers and investors; and
5. The acceptable level of transaction costs.

The three baseline scenarios are outlined in the following subsections. Each version involves the three quantification steps described above: (1) the baseline or the project reference case, (2) estimated project-related emission levels, and (3) net emission benefits of the project.

4.6.1 Emission Baselines: Version 1

The first scenario will be based on a static emission baseline. This means that the emissions are assumed to remain constant throughout the life of the project. This scenario does not take into consideration changes to vehicle emissions and equipment over time. In this case, the baseline emissions (i.e. the estimate of emissions absent the project) are assumed to be equal to the historic emissions of the gasoline vehicles prior to the project. Finally, Version 1 of the case study includes only tailpipe and refueling emissions. This is defined as vehicle operation.

Step 1: Historic Emissions

The historic emissions in this baseline scenario include relevant GHG emissions (CO_2 and CH_4) for the one-year prior to implementation of the project. In general, historic emissions should include data for at least 12 consecutive months prior to the project. Table 4-1 lists emission factors of CO_2 and CH_4.[153] The last column in the table lists emissions in terms of CO_2E. This means that emissions of CH_4 have been multiplied by 23 (the GWP for methane) to find the carbon dioxide equivalent global warming potential of methane. The resulting number (2.3) has then been added to the CO_2 emissions (410.0) to find the total emissions per mile for one conventional taxi (412.3 g CO_2E/mile). The contribution of N_2O is minimal and has been excluded throughout the case study.

Table 4-1. Historic Emissions 12 Months Prior to the Project

Emission Factors based on Vehicle Operation for One Conventional Taxi		
	Emissions: Grams/Mile	CO_2E Emissions: Grams/mile
CH_4	0.1	2.3 (0.1 x 23)
CO_2	410.0	+ 410.0
Total		__412.3__ grams CO_2E/mile

---

[153] The emission factors presented throughout the case study are hypothetical, and although they are similar to emissions factors from existing models, they are intentionally different to reflect the fact that different countries use different models and factors.

Emissions one year prior to project:

To calculate annual historic emissions, the CO_2E emission factor for vehicle operation is multiplied by the number of miles each taxi was driven for the previous year (80,000 miles) times the number of taxis (75). The product, which is divided by 1,000,000 to show the result in terms of metric tons, represents the historic emissions during the one year prior to the project.

412.3 g CO_2E/mile x 80,000 miles x 75 cars = 2,473.8 metric tons of CO_2E

Step 2: The Baseline or Reference Case

The reference case represents what would have happened if the GHG reduction project were not implemented. In this case, it is assumed that the 75 old gasoline taxis, with an average age of 8 years, would have remained on the road for the next 10 years (the life of the project). Because Version 1 of the case study assumes that emissions of the project are static, the GHG emissions of the fleet of taxis over the next 10 years will remain the same as current levels; that is, future emissions will equal the annual historic emissions of the representative taxi described in table 4-1 above. The only difference between Step 1 and Step 2 is that annual emissions are multiplied by 10 to derive project emissions over the life of the project.

Reference case – one year historic emissions multiplied by 10:

412.3 grams CO_2E/mile x 80,000 miles/year x 75 cars x 10 years = 24,738 metric tons of CO_2E over the life of the project

Step 3: The Project Case

The project case represents emissions with the project in place. In this instance, the project case refers to the emissions of the 75 CNG taxis over the 10-year life of the project. Table 4-2 presents the emissions factors of one new CNG taxi. As expected, CH_4 emissions of the CNG taxis are higher and CO_2 emissions are lower than the respective emissions of the gasoline taxis to be replaced. Again, the last column in the table lists the emissions factor in terms of CO_2 equivalent. This means that emissions of CH_4 have been multiplied by 23 (the GWP for methane) to find the carbon dioxide equivalent global warming potential of methane. Table 4-2 indicates that the emissions factor of one CNG taxi is 262.6 g of CO_2 equivalent per mile driven.

Table 4-2. Version 1 of Case Study – The Project Case

Vehicle Operation of One CNG Taxi		
	Emissions: Grams/mile	CO_2E Emissions: Grams/mile
CH_4	0.6	13.8 (0.6 x 23)
CO_2	250.0	+ 250.0
Total		263.8 g CO_2E/mile

To find emissions during the life of the project, the emissions factor of one CNG vehicle is multiplied by the average miles driven per year (80,000), the number of NGV vehicles replacing gasoline vehicles (75), and the expected number of years of the project (10).

Project case – emissions for one year multiplied by 10:

263.8 grams CO_2E/mile x 80,000 miles/year x 75 cars x 10 years = <u>15,828</u> metric tons of CO_2E over the life of the project

Step 4: Deriving Net Project Benefits

The net project emission benefits are derived by subtracting the project case from the reference case. As illustrated below, the net project benefits of Version 1 of the case study are 8,970 metric tons of CO_2 equivalent.

Reference case	-	project case	=	Net project benefits
24,726	-	15,756	=	<u>8,910</u> metric tons of CO_2E

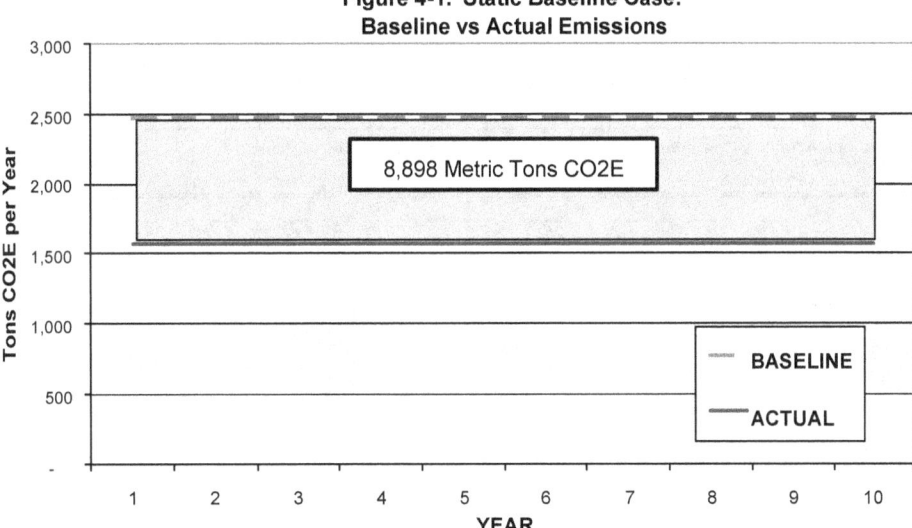

**Figure 4-1. Static Baseline Case:
Baseline vs Actual Emissions**

4.6.2 Emission Baselines: Version 2

The second scenario for the CNG vehicle project relies on a dynamic emission baseline. A dynamic baseline takes into account the changes that may happen to equipment, and thus emissions, as the vehicles age over time; that is, emissions of a vehicle will grow at an increasing rate every year. In this version, it will therefore no longer be sufficient to use the historic emissions as the reference point for the entire life of the project. Rather, the data for estimating the reference case and project emissions will have to be based on an evaluation of how the aging process (time dependent) influences both the conventional vehicles and the NGVs. Emissions will be affected by changes in engine efficiencies over time and the replacement of older vehicles with new ones. The numbers used in this case study are hypothetical and are not based on any particular studies on the relationship between emissions and vehicle age. As was the case with Version 1 of this case study, Version 2 includes only tailpipe and refueling emissions, or those associated with vehicle operations.

Step 1: The Baseline or Reference Case

The reference case represents what would have happened if the GHG reduction project were not implemented. As in Version 1 of this case study, it is assumed that the 75 old gasoline taxis, with an average age of 8 years, would have remained on the road for the next 10 years. However, in this second emission baseline scenario, it is assumed that the emissions of the old vehicles would have increased over time due to equipment failure and aging of the gasoline taxis. In addition, it is assumed that 8 vehicles (~10%) would have been replaced by new gasoline vehicles due to age or accidents, thereby slowing emissions growth. These changes are quantified in Table 4-3. Note that the initial emissions factors are based on the historical data presented in Version 1.

The first row in the table lists the estimated CH_4 emission factor (grams/mile) for one taxi expressed in terms of CO_2 equivalent. The second row lists the CO_2 emission factor during the life of the project. These emission factors are summed and listed in row three of the table. Finally, in the bottom row, we have multiplied the emission factor of one taxi (e.g.: 412.3 g CO_2E/mile for year one) by the average miles driven annually of one car (80,000) by the number of vehicles in the fleet (75) to find the emissions of the entire project for that year (and converted the total project emissions each year into metric tons). As shown in the last column of the table, the total emissions of the reference case (the sum of the annual emissions) are estimated to reach 25,495 metric tons of CO_2 equivalent during the 10-year life of the project.

Table 4-3. Version 2 of Case Study – The Reference Case

Emissions due to Vehicle Operation (Grams per Mile) -- Conventional Gasoline Vehicle										
	1	2	3	4	5	6	7	8	9	10	Total
CH_4 (in CO_2E*)	2.3	2.3	2.3	2.5	2.5	2.8	2.8	3.0	3.2	3.5	
CO_2	410.0	412.0	414.0	417.0	420.0	423.0	425.0	429.0	434.0	438.0	
Total gCO_2E/mile	412.3	414.3	416.3	419.5	422.5	425.8	427.8	432.0	437.2	441.5	
Total tCO_2E/year**	2,473.8	2,485.8	2,497.8	2,517.2	2,535.2	2,554.6	2,566.6	2,591.9	2,623.3	2,648.7	**25,495**

* CO_2E (CO_2 Equivalent) = CH_4 x 23

** Total metric tons CO_2E/year = (grams CO_2E/mile * 80,000 miles/car/year * 75 cars) / (1,000,000 grams/metric ton)

Step 2: The Project Case

The project case represents the expected emissions with the project in place. As in Version 1 of this case study, the project case refers to the emissions of the 75 CNG taxis over the 10-year life of the project. However, in this case it is assumed that emissions will increase over time due to equipment failure and aging. In addition we assume that 3 of the vehicles (~4%) would have been replaced by new NGVs due to age or accidents, slowing emissions growth. These changes are quantified in Table 4-4. The last row lists annual emissions in terms of metric tons of CO_2 equivalent.

The first row in the table lists the estimated CH_4 emission factor (grams/mile) for one taxi expressed in terms of CO_2 equivalent. The second row lists the CO_2 emission factor. These emission factors are summed and listed in row three of the table. Finally, in the bottom row, we have multiplied the emission factor of one taxi (e.g.: 252.8 g CO_2E for year one) times the average miles driven (80,000 miles/year) of one car. times the number of vehicles in the fleet (75) to find the emissions of the entire project for each year (and converted to units of metric tons). As shown in the last column of the table, the total emissions during the life of the project are projected to be 15,413 metric tons of CO_2 equivalent.

Table 4-4. Version 2 of Case Study – The Project Case

Emissions due to Vehicle Operation (Grams per Mile) -- CNG Vehicle											
	1	2	3	4	5	6	7	8	9	10	Total
CH_4(in CO_2E*)	2.8	2.8	3.1	3.1	3.1	3.3	3.3	3.5	3.8	4.0	
CO_2	250.0	250.0	251.0	251.0	252.0	253.0	254.0	256.0	258.0	261.0	
Total g CO_2E*/mile	252.8	252.8	254.1	254.1	255.1	256.3	257.3	259.5	261.8	265.0	
Total tCO_2E/year**	1,517.1	1,517.1	1,524.5	1,524.5	1,530.5	1,537.8	1,543.8	1,557.2	1,570.6	1,590.0	**15,413**

* CO_2E (or CO_2 Equivalent) = CH_4 x 23

** Total metric tons CO_2E/year = (grams CO_2E/mile * 80,000 miles/car/year * 75 cars) / (1,000,000 grams/metric tons)

Step 3: Deriving Net Project Benefits

The net project emission benefits are derived by subtracting the project case from the reference case. As illustrated below, the net project benefits of Version 2 of the case study are 10,085 metric tons of CO_2 equivalent.

Reference case   -   project case   =   Net project benefits

25,495   -   15,413   =   10,082 metric tons of CO_2E

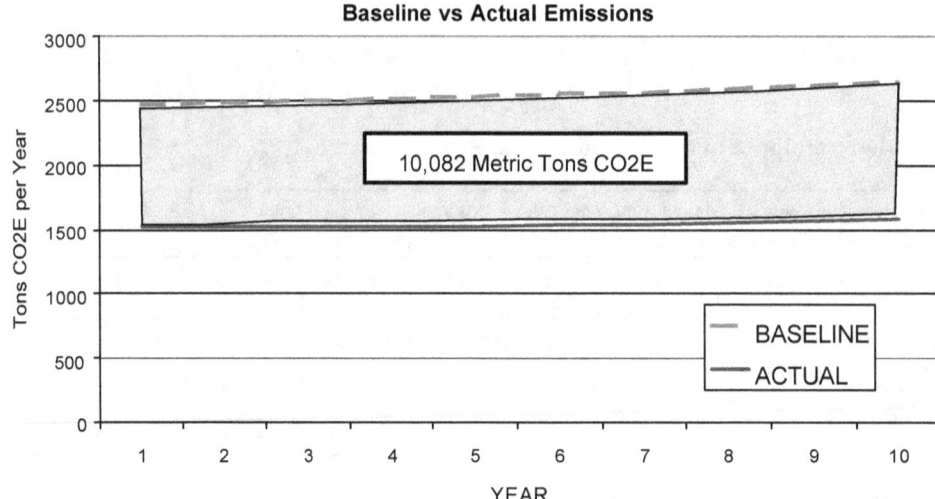

**Figure 4-2. Dynamic Baseline Case:
Baseline vs Actual Emissions**

4.6.3   Emission Baselines: Version 3

The third version of the emission baseline for the CNG vehicle project also uses a dynamic emission baseline. The baseline incorporates changes due to age and equipment failure over time. However, unlike Versions 1 and 2 of this case study, this emission baseline includes emissions from the entire fuel cycle of the CNG and gasoline vehicles. Hence, this baseline is much more detailed than the two previous versions. Emissions data are presented for three stages of the fuel cycle: feedstock, fuel and vehicle operation. [154] It is assumed that while emissions from the vehicle operation stage are time dependent as in Version 2, emissions from the fuel and feedstock stages are static, since they do not depend on the vehicles themselves.

Step 1: Historic Emissions

The historic emissions in this baseline scenario include relevant GHG emissions for the entire fuel cycle and are based on data collected as of one year prior to the implementation of the project. Table 4-5 lists emission factors of CO_2 and CH_4 for a single conventional gasoline taxi scheduled to be replaced by the project. The last column in the table lists emissions in terms of CO_2 equivalent. This means that emissions of CH_4 have been multiplied by 23 (the GWP for methane) to find the carbon dioxide equivalent global warming potential of methane.

---

[154] The feedstock-related stage includes feedstock recovery, transportation, and storage. The fuel-related stage includes fuel production, transportation, storage, and distribution. The vehicle operation stage includes vehicle refueling, tailpipe and operations.

Table 4-5. Version 3 of Case Study – Historic Emissions 12 Months Prior to the Project

Vehicle Operation of One Conventional Taxi				
	Feedstock	Fuel	Vehicle Operation	Total
	Emissions: Grams/mile	Emissions: Grams/mile	Emissions: Grams/mile	CO_2E Emissions: Grams/mile
CH_4	0.80	0.08	0.10	22.54 (0.98x23)
CO_2	30	75	410	+515
Total				**537.54 grams CO_2E/mile**

Emissions one year prior to project:

Once the CO_2 equivalent emission factor of one taxi has been derived (498.04 g CO_2 equivalent per mile), this number is multiplied times the average number of miles driven per year (80,000 miles) times the number of vehicles in the fleet (75 cars). The product is then converted to the final units of metric tons.

537.54 g CO_2E/mile x 80,000 miles x 75 cars = **3,225.24** metric tons of CO_2E

Step 2: The Baseline or Reference Case

The reference case represents what would have happened if the GHG reduction project were not implemented. As in the previous versions of this case study, it is assumed that the 75 old gasoline taxis, with an average age of 8 years, would have remained on the road for the next 10 years. In addition, it is assumed that the emissions of the old vehicles would have increased at a growing rate due to equipment failure and aging of the gasoline taxis, and that 8 of the vehicles (~10%) would have been replaced by new gasoline vehicles due to age or accidents, slowing emissions growth. Finally, data is collected for the entire fuel cycle of the project (Tables 4-6.A-C).[155]

---

[155] As with the GREET model we have divided the full fuel cycle intro three stages: the feedstock-related stage includes feedstock recovery, transportation and storage. The fuel-related stage includes fuel production, transportation, storage and distribution. The vehicle operation stage includes vehicle refueling, vehicle operations, and fuel combustion (also known as tailpipe emissions).

Table 4-6.A Version 3 of Case Study – The Reference Case (Feedstock)

Feedstock (grams/mile)											
	1	2	3	4	5	6	7	8	9	10	Total
CH_4 (in CO_2E^*)	11.5	11.5	11.5	11.5	11.5	11.5	11.5	11.5	11.5	11.5	
CO_2	30.0	30.0	30.0	30.0	30.0	30.0	30.0	30.0	30.0	30.0	
Total g CO_2E/mile	41.5	41.5	41.5	41.5	41.5	41.5	41.5	41.5	41.5	41.5	
Total tCO_2E/year**	249.0	249.0	249.0	249.0	249.0	249.0	249.0	249.0	249.0	249.0	2,490.0

Table 4-6.B Version 3 of Case Study – The Reference Case (Fuel)

Fuel (grams/mile)											
	1	2	3	4	5	6	7	8	9	10	Total
CH_4 (in CO_2E^*)	3.0	3.0	3.0	3.0	3.0	3.0	3.0	3.0	3.0	3.0	
CO_2	75.0	75.0	75.0	75.0	75.0	75.0	75.0	75.0	75.0	75.0	
Total g CO_2E/mile	78.0	78.0	78.0	78.0	78.0	78.0	78.0	78.0	78.0	78.0	
Total tCO_2E/year**	467.9	467.9	467.9	467.9	467.9	467.9	467.9	467.9	467.9	467.9	4,679.4

Table 4-6.C Version 3 of Case Study – The Reference Case (Vehicle Operation)

Vehicle Operation (grams/mile)											
	1	2	3	4	5	6	7	8	9	10	Total
CH_4 (in CO_2E^*)	2.3	2.3	2.3	2.5	2.5	2.8	2.8	3.0	3.2	3.5	
CO_2	410.0	412.0	414.0	417.0	420.0	423.0	425.0	429.0	434.0	438.0	
Total g CO_2E/mile	412.3	414.3	416.3	419.5	422.5	425.8	427.8	432.0	437.2	441.5	
Total tCO_2E/year**	2,473.8	2,485.8	2,497.8	2,517.2	2,535.2	2,554.6	2,566.6	2,591.9	2,623.3	2,648.7	25,494.8

* CO_2E (CO_2 Equivalent) = CH_4 x 23

** Total metric tons CO_2E/year = (grams CO_2E/mile * 80,000 miles/car/year * 75 cars) / (1,000,000 grams/metric ton)

The first row in Tables 4-6 A-C lists the estimated CH_4 emission factor (grams/mile) for one taxi expressed in terms of CO_2 equivalent. The second row lists the per-mile CO_2 emission factor. These emission factors are summed in row three of the tables. In the bottom row, we have multiplied the emission factor of one taxi with the average miles driven annually (80,000) of each car. The result is then multiplied by the number of vehicles in the fleet (75) to find the total emissions for each year, and then converted to metric tons of CO_2E.

Emissions from the entire fuel cycle for the entire life of the project (feedstock = 2,739 metric tons CO_2E; fuel = 4,823 metric tons CO_2E; vehicle operation = 25,494 metric tons CO_2E) are added together, resulting in emissions for the 10-year reference case of 33,054 metric tons of CO_2 equivalent.

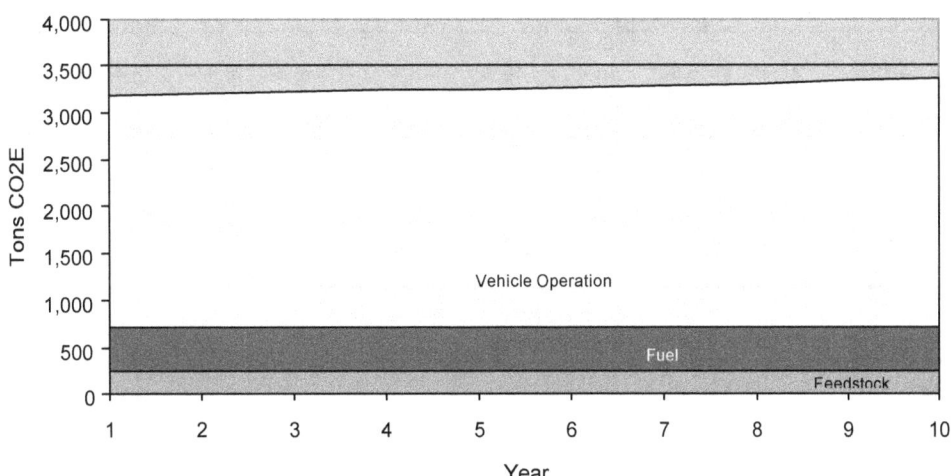

**Figure 4-3. Case Study Version 3 Baseline:
Annual GHG Emissions by Fuel Stage**

Step 3: The Project Case

As in the previous versions of this case study, the project case refers to the expected emissions of the 75 CNG taxis over the 10-year life of the project. It is assumed that emissions will increase at an accelerating rate due to equipment failure and aging. In addition, we assume that 3 of the vehicles (~4%) would have been replaced by new NGVs due to age or accidents, slowing emissions growth. However, in this version, data for the entire project fuel cycle is included in the analysis. These figures are presented in Tables 4-7 A-C, in the same fashion as in Tables 4-6 A-C above.

The emissions from the entire fuel cycle for the entire life of the project (feedstock = 2,946 metric tons CO_2E; fuel = 2,354 metric tons CO_2E; vehicle operation = 16,070 metric tons CO_2E) are added together, resulting in emissions of 21,371 metric tons of CO_2 equivalent.

Table 4-7.A Version 3 of Case Study – The Project Case (Feedstock)

Feedstock (grams/mile)											
	1	2	3	4	5	6	7	8	9	10	Total
CH_4(in CO_2E*)	18.4	18.4	18.4	18.4	18.4	18.4	18.4	18.4	18.4	18.4	
CO_2	28.0	28.0	28.0	28.0	28.0	28.0	28.0	28.0	28.0	28.0	
Total g CO_2E/mile	46.4	46.4	46.4	46.4	46.4	46.4	46.4	46.4	46.4	46.4	
Total tCO_2E/year**	278.4	278.4	278.4	278.4	278.4	278.4	278.4	278.4	278.4	278.4	**2,784.0**

Table 4-7.B Version 3 of Case Study – The Project Case (Fuel)

Fuel (grams/mile)											
	1	2	3	4	5	6	7	8	9	10	Total
CH_4(in CO_2E*)	1.8	1.8	1.8	1.8	1.8	1.8	1.8	1.8	1.8	1.8	
CO_2	35.0	35.0	35.0	35.0	35.0	35.0	35.0	35.0	35.0	35.0	
Total g CO_2E/mile	36.8	36.8	36.8	36.8	36.8	36.8	36.8	36.8	36.8	36.8	
Total tCO_2E/year**	221.0	221.0	221.0	221.0	221.0	221.0	221.0	221.0	221.0	221.0	**2,210.4**

Table 4-7.C Version 3 of Case Study – The Project Case (Vehicle Operation)

Vehicle Operation (grams/mile)											
	1	2	3	4	5	6	7	8	9	10	Total
CH_4(in CO_2E*)	13.8	13.8	14.0	14.0	14.0	14.3	14.3	14.5	14.7	15.0	
CO_2	250.0	250.0	251.0	251.0	252.0	253.0	254.0	256.0	258.0	261.0	
Total g CO_2E/mile	263.8	263.8	265.0	265.0	266.0	267.3	268.3	270.5	272.7	276.0	
Total tCO_2E/year**	1,582.8	1,582.8	1,590.2	1,590.2	1,596.2	1,603.6	1,609.6	1,622.9	1,636.3	1,655.7	**16,070.2**

* CO_2E (CO_2 Equivalent) = CH_4 x 23

** Total metric tons CO_2E/year = (grams CO_2E/mile * 80,000 miles/car/year * 75 cars) / (1,000,000 grams/metric ton)

Figure 4-4. Case Study Version 3 Project Emissions:
Annual GHG Emissions by Fuel Stage

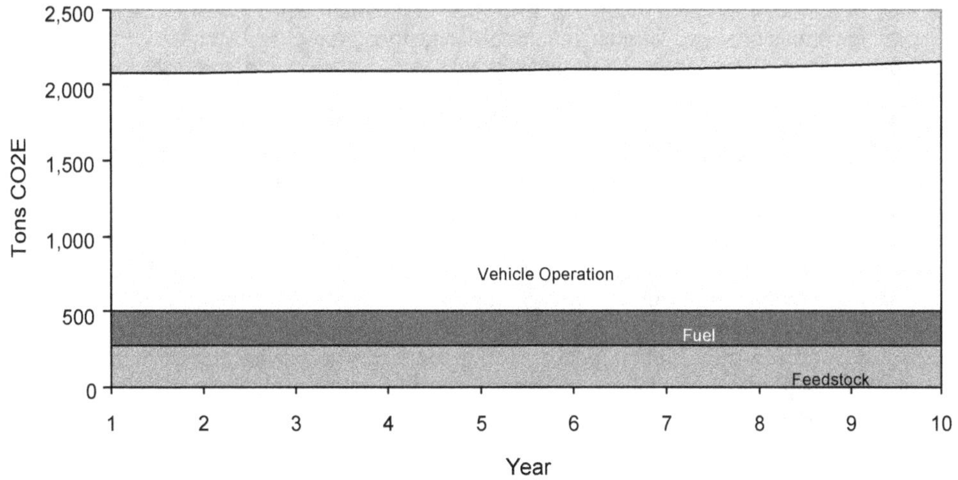

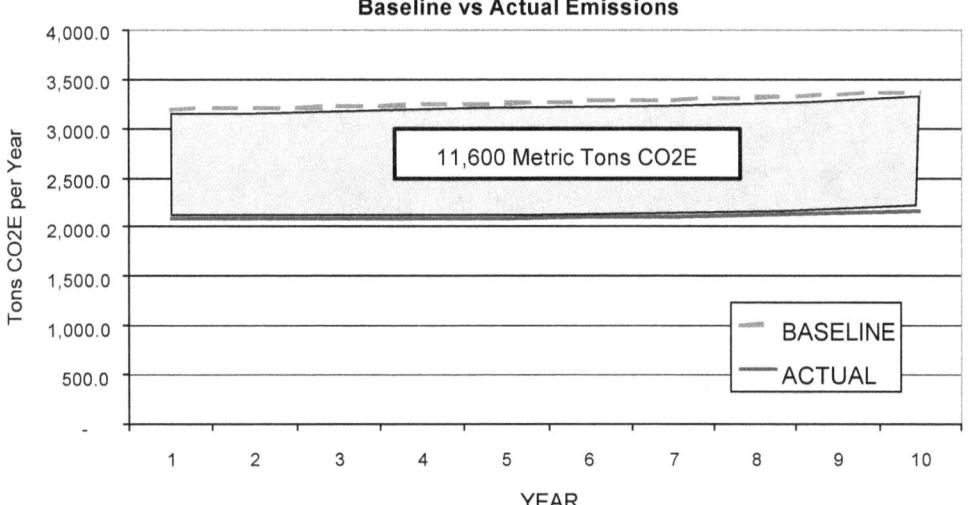

Step 4: Deriving Net Project Benefits

The net project emission benefits are derived by subtracting the project case from the reference case. As illustrated below, the net project benefits of Version 3 of the case study are 11,776 metric tons of CO_2 equivalent.

Reference case	-	project case	=	Net project benefits
33,054	-	21,377	=	11,687 metric tons of CO_2E

Figure 4-5. Dynamic Baseline Case; Full Fuel Cycle:
Baseline vs Actual Emissions

4.7   Discussion

The previous section presented three different methods for estimating the emission benefits of an NGV project. Each of the three baselines represents a viable means of calculating emissions reductions resulting from the project. Ultimately, the specific circumstances of a potential NGV project will determine which methodology is being used.

In general, there is a tradeoff between the accuracy of a baseline and the cost and effort associated with its calculation. As transportation projects are typically smaller in size and thus reap relatively few GHG emission reduction credits, it will be harder to justify the transaction costs involved with developing a very detailed estimate of projected emissions. As a result, project developers may prefer to use a less stringent baseline estimation procedure. However, as a general rule, project developers should aim to be as conservative as possible when determining emission reduction credits. Investors looking to purchase emission reduction credits want to ensure that the credits purchased are credible (additional, verifiable, transparent) and minimize the risk of default in future commitment periods. Hence, they prefer to invest in credits that are based on sound and credible estimation procedures. It is therefore important that project developers clearly describe the baseline methodology and assumptions used, and explain why this approach was preferred over other methods. Moreover, in cases where there might be some uncertainty regarding the exact amount of expected emissions benefits—for example due to an expected decline in NGV efficiency which cannot yet be quantified because of little experience with the technology—project developers should select the least optimistic emissions scenario. This type of estimation procedure is much more likely to gain acceptance by current and future GHG crediting programs.

5 Summary and Conclusions

Accurate and verifiable emission reductions are a function of the degree of transparency and stringency of the protocols employed in documenting project- or program-associated emissions reductions. The purpose of this guide is to provide a background for law and policy makers, urban planners, and project developers working with the many GHG emission reduction programs throughout the world to quantify and/or evaluate the GHG impacts of NGVs.

In order to evaluate the GHG benefits and/or penalties of NGV projects, it is necessary to first gain a fundamental understanding of the technology employed and the operating characteristics of these vehicles, especially with regard to the manner in which they compare to similar conventional gasoline or diesel vehicles. Therefore, the first two sections of this paper explain the basic technology and functionality of NGVs, but focus on evaluating the models that are currently on the market with their similar conventional counterparts, including characteristics such as cost, performance, efficiency, environmental attributes, and range.

Since the increased use of NGVs, along with AFVs in general, represents a public good with many social benefits at the local, national, and global levels, NGVs often receive significant attention in the form of legislative and programmatic support. Some states mandate the use of NGVs, while others provide financial incentives to promote their procurement and use. Furthermore, Federal legislation in the form of tax incentives or procurement requirements can have a significant impact on the NGV market. In order to implement effective legislation or programs, it is vital to have an understanding of the different programs and activities that already exist so that a new project focusing on GHG emission reduction can successfully interact with and build on the experience and lessons learned of those that preceded it.

Finally, most programs that deal with passenger vehicles—and with transportation in general—do not address the climate change component explicitly, and thus there are few GHG reduction goals that are included in these programs. Furthermore, there are relatively few protocols that exist for accounting for the GHG emissions reductions that arise from transportation and, specifically, passenger vehicle projects and programs. These accounting procedures and principles gain increased importance when a project developer wishes to document in a credible manner, the GHG reductions that are achieved by a given project or program. Section four of this paper outlined the GHG emissions associated with NGVs, both upstream and downstream, and section five illustrated the methodology, via hypothetical case studies, for measuring these reductions using different types of baselines.

Unlike stationary energy combustion, GHG emissions from transportation activities, including NGV projects, come from dispersed sources creating a need for different methodologies for assessing GHG impacts. This resource guide has outlined the necessary context and background for those parties wishing to evaluate projects and develop programs, policies, projects, and legislation aimed at the promotion of NGVs for GHG emission reduction.

Appendices

A1 U.S. State and Local Climate Change Legislation and Policy Initiatives

States and local governments have taken up a variety of actions that seek to enable governments and private stakeholders begin to implement the basic tools to prepare for long-term regulatory responses to climate change. Such actions include preparing state climate change action plans, establishing state GHG emissions reporting registries, launching pilot trading programs in GHG emissions credits. In addition, a number of state and local "multi-pollutant" legislation include provisions for setting limits and trading opportunities for various air pollutants that include carbon dioxide.

This appendix provides a summary of 18 recent state, local, and regional legislation and policy initiatives that relate to efforts to improve GHG emissions reporting and, potentially, trading. It should be noted that a number of policy actions have not specifically been the product of legislation, but rather a result of policy and program measures adopted by state government bodies. While these initiatives have not been included in the summary below, they are making a relevant impact on state and local efforts to address climate change.

Much of the motivation in states to develop such actions had been prompted in the late 1990s by the lack of serious action by the Federal government, and in many cases this has led to the creation of a patchwork of standards and requirements from state to state. However, on February 14, 2002, President Bush announced the Administration's official policy on climate change:

> Our immediate goal is to reduce America's GHG emissions relative to the size of our economy... Our government will also move forward immediately to create world-class standards for measuring and registering emission reductions. And we will give transferable credits to companies that can show real emission reductions.[156]

The President's "Global Climate Change Policy Book" specifically addresses local and national GHG registries:

> The President directed the Secretary of Energy, in consultation with the Secretary of Commerce, the Secretary of Agriculture, and the Administrator of the EPA, to propose improvements to the current voluntary emission reduction registration program under section 1605(b) of the 1992 Energy Policy Act within 120 days... A number of proposals to reform the existing registry—or create a new registry—have appeared in energy and/or climate policy bills introduced in the past year. The Administration will fully explore the extent to which the

---

[156] White House Office of the Press Secretary, "President Announces Clear Skies & Global Climate Change Initiatives," February 14, 2002, http://www.whitehouse.gov/news/releases/2002/02/20020214-5.html.

existing authority under the Energy Policy Act is adequate to achieve these reforms.[157]

Many states reacted to the Administration's policy and were eager to respond to the recommendations expected after 120 days. Project developers should keep abreast of developments under the 1605(b) program and other current events with respect to similar climate change legislation and policy initiative at the state and local level that may affect the emerging regulatory framework and budding market in GHG emissions.

---

[157] White House, Global Climate Change Policy Book, February 2002, http://www.whitehouse.gov/news/releases/2002/02/climatechange.html.

Table A1. U.S. State and Local Climate Change Legislation and Policy Initiatives

Region/ State/ City	Directive	Date	Objective	Contact
	Senate Bill 1771	Signed September 30, 2000	SB 1771 (as amended by SB 527) established the California Climate Action Registry (California Registry), a non-profit organization designed to help California entities to participate in voluntary GHG emissions reporting, certification, and registration. Organizations located outside the State of California may also participate in the Registry. Under the law, the State of California will use its best efforts to ensure that organizations that voluntarily inventory their emissions receive appropriate consideration under any future international, federal, or state regulatory regimes relating to GHG emissions. SB 527 requires the California Energy Commission to provide guidance to the Registry on a number of issues, such as developing GHG emissions reporting protocols, qualifying third-party organizations to provide technical assistance, and qualifying third-party organizations to provide certification of emissions baselines and inventories.[158] The California Registry became operational in 2002 and prepared its reporting and certification protocols for release in late 2002.	California Climate Action Registry, Ann Hewitt, tel. 213-891-1444 or email **ann@climateregistry.org.**
California *	Senate Bill 527	Signed October 13, 2001		
	Assembly Bill 1493	Signed July 22, 2002	AB 1493 requires the California Air Resources Board (CARB) to adopt regulations for CO_2 emissions from passenger cars, light trucks, and SUVs by January 1, 2005. The bill directs the CARB to adopt regulations "that achieve the maximum feasible reduction of GHGs emitted by passenger vehicles and light-duty trucks and any other vehicles" in the state. The law would take effect January 1, 2006 and would apply to vehicles manufactured in MY2009 and later. In preparing regulations, CARB may not: (1) impose additional fees or taxes on motor vehicles, fuel, or miles traveled; (2) ban the sale of any vehicle category in the state; (3) require reductions in vehicle weight; (4) limit speed limits; or (5) limit vehicle miles traveled. AB 1493 would also require the California Climate Action Registry to develop reporting procedures by July 1, 2003, in consultation with CARB.[159]	Chuck Shulock (CARB), tel. (916) 322-6964, or email **cshulock@arb.ca.gov.**
Illinois	Senate Bill 372	Signed August 7, 2001	This multi-pollutant legislation requires the Illinois EPA to establish an interstate nitrogen oxide (NO_X) trading program and issue findings that address the need to control or reduce emissions of NO_X, sulfur dioxide (SO_2), mercury (Hg), and GHGs from fossil fuel-fired electric generating plants. The findings are to address the establishment of a banking system, consistent with DOE's Voluntary Reporting of Greenhouse Gases (1605(b)) Program for certifying credits for voluntary offsets of GHG emissions or reductions, of GHGs.[160]	For more information, contact Steven King at the Illinois Environmental Protection Agency, tel. 217-524-4792, or email **steven.king@epa.state.il.us**.

[158] California Climate Action Registry, **http://www.climateregistry.org.**
[159] State of California, Assembly Bill 1493 (July 22, 2002), **http://www.arb.ca.gov/gcc/gcc.htm.**
[160] State of Illinois, Senate Bill 372 (August 7, 2001), **http://www.legis.state.il.us.**

Table A1.   U.S. State and Local Climate Change Legislation and Policy Initiatives

Region/ State/ City	Directive	Date	Objective	Contact
Maine	Legislative Document 87	Passed April 6, 2001	This requires the Maine Department of Environmental Protection to develop rules to create a voluntary registry of GHG emissions. The rules must provide for the collection of data on the origin of the carbon emissions as either fossil fuel or renewable resources, and the collection of data on production activity to allow the tracking of future emission trends.[161]	Maine Department of Environmental Protection, tel. 800-452-1942.
Maryland	Executive Order 01.01.2001.02, "Sustaining Maryland's Future with Clean Power, Green Buildings and Energy Efficiency"	Signed March 13, 2001	This Executive Order states, "the [Maryland Green Buildings] Council shall develop a comprehensive set of initiatives known as the 'Maryland Greenhouse Gas Reduction Action Plan'; and The Council shall report annually to the Governor and to the General Assembly on the efforts of State agencies in the implementation of... the Greenhouse Gas Reduction Plan, and other energy efficiency, energy production and sustainability issues or policies the Council may have considered.[162] A November 2001 report by the Council states that goals for GHG reductions in Maryland "will be set for both the State facilities and operations as well as statewide reduction goals to be achieved through voluntary initiatives, policies, and programs."[163]	Maryland Green Buildings Council, **http://www.dgs.state.md.us/GreenBuildings/default.htm**, or Geraldine Nicholson, Maryland Energy Administration, tel. 410-260-7207, or e-mail **gnicholson@energy.state.md.us** .
Massachusetts	Department of Environmental Protection Regulation 310 CMR 7.29	Issued April 23, 2001	This rule requires the six highest-polluting power plants in Massachusetts to meet overall emission limits for NO_X and SO_2 by October 1, 2004 and begin immediate monitoring and reporting of mercury emissions. For the six affected plants, the rule caps total CO_2 emissions and creates an emission standard of 1,800 pounds of CO_2 per megawatt-hour (a reduction of 10 percent below the current average emissions rate). The CO_2 limits must be met by October 1, 2006 or October 1, 2008 for plant retrofits or replacements. Plant operators may meet the standard either by increasing efficiency at the plant, or by purchasing credits from other reduction programs approved by the Massachusetts Department of Environmental Protection.[164]	Massachusetts Department of Environmental Protection InfoLine, tel. 617-338-2255 or 800-462-0444, or email **dep.infoline@state.ma.us** ; or for Emissions Trading, contact Bill Lamkin, tel. 978-661-7657 or email **Bill.Lamkin@state.ma.us**, or for the Air Program Planning Unit that covers these regulations, see Nancy Seidman, tel. 617-556-1020, or email **Nancy.Seidman@state.ma.us**.

[161] State of Maine, Legislative Document 87 (April 6, 2001), http://janus.state.me.us/legis/bills/.

[162] State of Maryland, Executive Order 01.01.2001.02, http://www.gov.state.md.us/gov/execords/2001/html/0002eo.html.

[163] Maryland Green Buildings Council, "2001 Green Buildings Council Report," November 2001, pg. 30, http://www.dgs.state.md.us/GreenBuildings/Documents/FullReport.pdf.

[164] State of Massachusetts, 310 CMR 7.29. See U.S. Environmental Protection Agency, Legislative Initiatives, http://yosemite.epa.gov/globalwarming/ghg.nsf/actions/LegislativeInitiatives.

Table A1. U.S. State and Local Climate Change Legislation and Policy Initiatives

Region/ State/ City	Directive	Date	Objective	Contact
Michigan	Senate Bill 693	Introduced October 2001	This bill to amend the 1994 Natural Resources and Environmental Protection Act calls for declining caps in nitrogen oxides, SO_2, CO_2, and mercury by 2007.[165] The bill has been referred to the Committee on Natural Resources and Environmental Affairs.[166]	Office of Senator Alma Wheeler Smith (bill sponsor), tel. 800-344-2562 or 517-373-2406 or email **SenASmith@senate.state.mi.us**.
New England Governors/ Eastern Canadian Premiers	Climate Action Plan	Signed August, 2001	The New England Governors/Eastern Canadian Premiers' climate change action plan defines incremental goals for the coalition as follows: ■ in the short-term, reduce regional GHG emissions to 1990 emissions by 2010; ■ for the mid-term, reduce regional GHG emissions by at least 10 percent below 1990 emissions by 2020, and establish an iterative five-year process, beginning in 2005, to adjust existing goals, if necessary, and set future emissions reduction goals; and ■ for the long-term, reduce regional GHG emissions sufficiently to eliminate any dangerous threat to climate—current science suggests this will require reductions of 75–85 percent below current levels. The action plan also calls for the creation of a regional emissions registry and the exploration of a trading mechanism.[167]	New England Secretariat, New England Governors' Conference Inc., tel. 617-423-6900 or email **negc@tiac.net**
New Hampshire	House Bill 284, "Clean Power Act"	Approved January 2, 2002	This four-pollutant bill was the first in the nation to include CO_2.[168] Emission reduction requirements include 75% of sulfur dioxide by 2006; 70% of nitrogen oxide by 2006; 3% of carbon dioxide by 2006 (1990 levels); and mercury levels are still to be determined by 2004.[169]	New Hampshire Office of the Governor, tel. 603-271-2121.

[165] Jones, Brian M., "Emerging State and Regional GHG Emission Trading Drivers," presented at the Electric Utilities Environmental Conference, Tuscan, Arizona, January 2002.
[166] Michigan State Legislature, Senate Bill 0693, http://www.mileg.org.
[167] New England Governors/ Eastern Canadian Premiers, "Climate Change Action Plan," http://www.cmp.ca/CCAPe.pdf.
[168] New Hampshire, Office of the Governor, Press Releases, "Governor Shaheen Hails House Passage of Clean Power Act," http://www.state.nh.us/governor/media/010202clean.html.
[169] Jones, Brian M., "Emerging State and Regional GHG Emission Trading Drivers," presented at the Electric Utilities Environmental Conference, Tuscan, Arizona, January 2002.

Table A1. U.S. State and Local Climate Change Legislation and Policy Initiatives

Region/ State/ City	Directive	Date	Objective	Contact
	Senate Bill 159	Approved July 6, 1999	This bill established a registry for voluntary GHG emission reductions to create an incentive for voluntary emission reductions.[170] Implementation rules were adopted on February 23, 2001 under the New Hampshire Code of Administrative Rules, Chapter Env-A 3800 (Voluntary Greenhouse Gas Emissions Reductions Registry).[171]	Joanna Morin, Department of Environmental Science, tel. 800-498-6868 or 603-271-1370, or email **jmorin@desstate.nh.us**.
New Jersey	N.J.A.C 7:27-30.2 and 30.5	Adopted April 17, 2000	The New Jersey Department of Environmental Protection recently adopted rules to add provisions to the Open Market Emissions Trading Program for the generation and banking of GHG credits. The GHGs included are: CO_2; CH_4; nitrous oxide (N_2O); certain hydro fluorocarbons, per fluorocarbons; and sulfur hexafluoride. The Program was established to provide incentives for voluntary reduction of air contaminant emissions and also provide an alternative means for regulated entities to achieve compliance with air pollution control obligations in a more cost-effective manor.[172] The draft guidance on the preparation of quantification protocols is provided at **www.state.nj.us/dep/aqm/omet/**.	New Jersey Department of Environmental Protection, Air Quality Management Bureau of Regulatory Development, tel. 609-777-1345 or email **aqrdweb@dep.state.nj.us**.
New York State	Executive Order (Greenhouse Gas Task Force)	Created June, 2001	New York Governor Pataki established a Greenhouse Gas Task Force to prepare policy recommendations on climate change.[173] Preliminary recommendations for actions and policies from the Task Force's Working Groups included establishing a statewide target for GHG emission reductions relative to 1990 levels, and establishing a greenhouse registry to document baseline emissions and voluntary emissions reductions for participating customers. The Task Force plans a Final Report to be complete by March 2002.[174]	See the Governor's website at **www.state.ny.us/governor**
	Assembly Bill 5577	Introduced February 27, 2001	This multi-pollutant bill provides for regulation of emissions of nitrogen oxide, SO_2 and CO_2. The bill passed the Assembly on March 25, 2002, and was referred to the Senate Environmental Conservation Committee.[175]	For more information, contact the bill sponsor, Richard Brodsky, tel. 518-455-5753 or 914-345-0432, or email **brodskr@assembly.state.ny.us**.

---

[170] New Hampshire, Senate Bill 0159, **http://www.gencourt.state.nh.us/legislation/1999/SB0159.html**.
[171] New Hampshire Department of Environmental Services, **http://www.des.state.nh.us/ard/climatechange/ghgr.htm**.
[172] New Jersey, Department of Environmental Protection, Air Quality Permitting, Air Quality Management, Air And Environmental Quality Enforcement, "Open Market Emissions Trading Rule," **http://www.state.nj.us/dep/aqm/ometp2ad.htm**.
[173] Press Release, Office of the Governor of New York State, June 10, 2001, **http://www.state.ny.us/governor/press/year01/june10_01.html**.
[174] New York State, Draft State Energy Plan, December 2001, **http://www.nyserda.org/draftsepsec2.pdf**.
[175] New York State Assembly, Bill 5577, **http://www.assembly.state.ny.us/leg/?bn=A.5577**.

Table A1.   U.S. State and Local Climate Change Legislation and Policy Initiatives

Region/ State/ City	Directive	Date	Objective	Contact
New York City	New York City Council Bill No. 30[176]	Reintroduced January 30, 2002	New York City Council Member Peter Vallone, Jr. reintroduced a bill that would require the city's power plants to reduce CO_2 emissions by as much as 20 percent within five years of enactment. Under the terms of the legislation, owners of power plants that produce at least 25 megawatts of electricity for sale would be required to pay high fines for generators that emit levels of CO_2 that exceed those established by an independent board.[177]	New York City Council Member Peter Vallone Jr. tel. 718 274-4500 or 212-788-6963, or email vallonejr@council.nyc.ny.us .
Suffolk County	Carbon Dioxide Law	Passed July 24, 2001	New York State's Suffolk County became the first county to pass a resolution limiting CO_2 emissions. The resolution seeks to encourage efficiency in existing power plants and future facilities by setting allowable rates for CO_2 emissions and penalties for exceeding those limits. Under the law, starting on March 1, 2002 any power plant in the county that generates over 1,800 pounds of CO_2 emissions per Megawatt-hour would be fined two dollars for every ton above the limit. An additional $1 per excess ton would be charged in each consecutive year. The bill contains several alternatives to paying fines including buying emission credits through nationally recognized CO_2 trading markets, investing in alternative energy sources, or donating penalties to community environmental groups.[178]	Suffolk County Executive's Office, tel. 631-853-4000.
Nassau County	Carbon Dioxide Proposal	N/A	Nassau County has considered proposing a CO_2 emissions intensity standard of 1,800 lbs/MWh, along with an allowable county-wide emission rate reduction of one percent for every 100 MW of electric generating capacity installed within the County until county emission are reduced by 20 percent. Under the proposed scheme, compliance could be met through emissions trading. Penalties of $2/ton in the first year and $1/ton each consecutive year would be assessed for non-compliance.[179]	Nassau County Government at tel. 516-571-3000.
North Carolina	Senate Bill 1078 (House Bill 1015), "Clean Smokestack Bill"	Passed the Senate on April 23, 2001	This bill would reduce emissions of nitrogen oxides by 78 percent by 2009 and sulfur dioxide by 73 percent by 2013. The bill also directs the North Carolina Division of Air Quality (DAQ) to study the issues for standards of reductions of mercury and carbon dioxide. DAQ is also planning to develop and adopt a	North Carolina Division of Air Quality at tel. 919-733-3340, or for Climate Change and Greenhouse Gases, contact Russell Hageman

[176] New York City Council Bill Int. No. 30, http://www.council.nyc.ny.us/pdf_files/bills/int0030-2002.htm.

[177] Forbes, "NYC Council Seeks Cut In Power Plant CO2 Emissions," January 30, 2002, www.forbes.com/newswire/2002/01/30/rtr498771.html.

[178] Suffolk County, Press Release, "Suffolk Becomes First County to Limit CO₂ Emissions," July 24, 2001, http://www.co.suffolk.ny.us/exec/press/2001/emissions.html.

[179] Jones, Brian M., "Emerging State and Regional GHG Emission Trading Drivers," presented at the Electric Utilities Environmental Conference, Tuscan, Arizona, January 2002.

Table A1. U.S. State and Local Climate Change Legislation and Policy Initiatives

Region/ State/ City	Directive	Date	Objective	Contact
			program of incentives to promote voluntary reductions of emissions including, emissions banking and trading and credit for voluntary early action. The bill was sent by the Senate to the House where is has been referred the Committee on Public Utilities [180]	at tel. 919-733-1490 or email **Russell.Hageman@ncmail.net**, Jill Vitas at tel. 919-715-8666 or email **Jill.Vitas@ncmail.net**
Oregon	House Bill 3283	Signed June 26, 1997	This bill established a carbon dioxide standard requiring new power generators to emit 17 percent less than the most energy efficient plant available. [181] The bill capped CO_2 emissions at 0.7 pounds of CO_2 per kilowatt-hour for base-load natural gas-fired power plants; in 1999 the cap was lowered to 0.675 pounds per kilowatt-hour. New energy facilities built in the state must avoid, sequester, or pay a per-ton of CO_2 offset into the Oregon Climate Trust. [182] The nonprofit Oregon Climate Trust accepts mitigation funds from energy facilities for displacing their unmet emissions requirements, and in turn must use the funds to carry out projects that avoid, sequester, or displace the CO_2. In January 2001, the Climate Trust released a request for proposals (RFP) to fund $5.5 million in carbon dioxide mitigation projects. [183]	Oregon Climate Trust, Mike Burnett, Executive Director, tel. 503-238-1915 or email: **info@climatetrust.org**. See also **www.climatetrust.org**.
Texas	TNRCC Report on Greenhouse Gases and Recommendations from the Executive Director	Presented January 18, 2002	In August 2000, the Texas Natural Resources Conservation Commission (TNRCC) issued a decision instructing the agency's Executive Director to prepare a report on GHGs. The draft report and recommendations from the Executive Director were presented to TNRCC commissioners at a public work session on January 18, 2002. The report included recommendations to "[develop] and maintain a voluntary registry for reporting GHG emission reductions resulting from specific emission reduction or sequestration projects and energy efficiency improvements within Texas. [184]	Texas Office of Environmental Policy, Analysis and Assessment, tel. 512-239-4900, or email **policy@tnrcc.state.tx.us** .

[180] North Carolina General Assembly, Senate Bill 1078 (House Bill 1015), http://www.ncga.state.nc.us/gascripts/billnumber/billnumber.pl?Session=2001&BillID=S1078.

[181] Oregon House Bill 3283, http://www.leg.state.or.us/97reg/measures/hb3200.dir/hb3283.int.html.

[182] U.S. Environmental Protection Agency, Legislative Initiatives, http://yosemite.epa.gov/globalwarming/ghg.nsf/actions/LegislativeInitiatives.

[183] See Oregon Climate Trust, http://www.climatetrust.org.

[184] The Chairman directed staff, before executing this recommendation, to evaluate the DOE 1605(b) voluntary greenhouse gas registry program, as is or with some changes, as a possible element of a Texas GHG registry which avoids duplicative reporting. See Texas Natural Resource Conservation Commission, Office of Environmental Policy, Analysis and Assessment, "Overview and Recommendations Identified by A Report to the Commission on Greenhouse Gases," February 8, 2002, http://www.tnrcc.state.tx.us/oprd/sips/greenhouse/.

Table A1. U.S. State and Local Climate Change Legislation and Policy Initiatives

Region/ State/ City	Directive	Date	Objective	Contact
Washington	Senate Bill 5674	Passed the House on March 13, 2001; Reintroduced to Senate	Senate Bill 5674 was passed by the Washington State House of Representatives on March 13, 2001 and was referred on motion to the Senate Environment, Energy & Water Committee on January 16, 2002.[185] This bill authorizes the establishment of an independent, nonprofit organization known as the Washington Climate Center to serve as a central clearinghouse for all climate change activities in the state. The Climate Center's activities include determining current and projected GHG emissions in the state, and studying and recommending the most cost-effective methods for reducing all net GHG emissions.[186]	Bill sponsors: Sen. Ken Jacobsen tel. 360-86-7690 or email **jacobsen_ke@leg.wa.gov**; Sen. Margarita Prentice tel. 360-786-7616 or email **prentice_ma@leg.wa.gov**; Sen. Karen Fraser tel. 360-786-7642 or email **fraser_ka@leg.wa.gov**; Sen. Jeanne Kohl-Welles tel. 360-786-7670 or email **kohl_je@leg.wa.gov**, or former Sen. Dow Constantine tel. 206-296-1008 or email **dow.constantine@metrokc.gov**.
Wisconsin	Assembly Bill 627	February 8, 2000	This multi-pollutant bill requires the Department of Natural Resources to establish and operate a system for registering reductions in emissions of GHGs if the reductions are required by law. The bill also authorizes the Department of Natural Resources to establish systems for registering reductions in fine particulate matter, mercury, and other air contaminants.[187]	Wisconsin Voluntary Emissions Reductions Registry Advisory Committee, **http://www.dnr.state.wi.us/org/a w/air/hot/climchgcom/**.
	Department of Natural Resources Rule NR 437	January 2002 (final draft approved for public hearing)	NR 437 is proposed to establish voluntary emissions reduction registries for various pollutants, including GHGs, mercury, fine particulate matter, and other contaminants. The rule represents a new Department of Natural Resources policy to systematically record and track voluntary emission reductions by industries, electric utility companies, agricultural and forestry interests, and transportation and energy efficiency interests. NR 437 establishes the rules and procedures under which the new registry will operate. The rule also identifies the sources that are eligible to register reductions for GHGs like carbon dioxide, methane, nitrous oxide, hydrofluorocarbons, perfluorocarbons and sulfur hexafluoride, as well as for nitrogen oxides, sulfur dioxide, volatile organic compounds, carbon monoxide, mercury, lead and fine particulate matter.[188]	Eric Mosher tel. 608-266-3010, or e-mail **moshee@dnr.state.wi.us**.

[185] Washington Senate Bill 5674, http://www.leg.wa.gov/wsladm/billinfo/dspBillSummary.cfm?billnumber=5674.
[186] U.S. Environmental Protection Agency, Legislative Initiatives, http://yosemite.epa.gov/globalwarming/ghg.nsf/actions/LegislativeInitiatives.
[187] Wisconsin Assembly Bill 627, http://www.legis.state.wi.us/1999/data/AB627.pdf.
[188] Wisconsin Voluntary Emission Reductions Registry Advisory Committee, http://www.dnr.state.wi.us/org/aw/air/hot/climchgcom/.

A2 Natural Gas Vehicle Year 2000 Projects Reported to the U.S. Voluntary Reporting of Greenhouse Gases Program (1605(b))

Table A2-1 Natural Gas Vehicle Year 2000 Projects Reported to the U.S. Voluntary Reporting of Greenhouse Gases Program

Reporting Entity	Project Name	Project Size *	Reported CO_2 Equivalent Reduction in 2000 (metric tons)
Baltimore Gas & Electric Company			
Alternatively Fueled Vehicles		163 vehicles	Direct: 134.08
Project Description	Operation of various numbers of Alternatively-Fueled Vehicles using Compressed Natural Gas.		
Estimation Method	CO_2 comparisons are based upon DOE data indicating that the CO_2 emission coefficient for gasoline is 156.7 pounds of CO_2 per million BTU and the coefficient for natural gas is 117.1 pounds of CO_2 per million BTU (DOE EIA-1605(1998)). Assumed vehicles travel 15,000 miles per year and gasoline has a heating value of 115,400 btu/mmBTU in an automotive application. Motor-gasoline vehicles have a fuel efficiency of approximately 288 mi/mmBTU and CNG vehicles have a fuel efficiency of 218 mi/mmBTU. Emissions are claimed for the CNG fuel consumed and reductions are claimed for the net of displaced motor gasoline emissions and emissions from the CNG fuel consumed.		
Central Hudson Gas & Electric Corporation			
Natural Gas Vehicles		4 vehicles	Direct: 8.88 Indirect: 12.13
Project Description	In 1988, several company fleet vehicles were converted to operate on natural gas with the ability to operate on gasoline retained. These conversions cost approximately $3,500 per vehicle. A 60cfm CNG station that was constructed to refuel these vehicles is still in operation. As the project continued, several more vehicles, mostly cars and light pick-up trucks, were converted, and new, factory-built NGVs were also purchased. Currently, a total of four (4) dedicated NG		

Estimation Method	vehicles (all Dodges) are operated by the Company. These vehicles include two (2) full-size and two (2) mini-vans. The meters located at the sole NGV refueling site, record both the amount of natural gas delivered, plus the gasoline gallon-equivalent (a roughly 8.415 gallons/1,000 cubic feet conversion). The CO2 emission rate (from Appendix B in the Form EIA-1605 Instructions) for natural gas (120.593 lbs CO2/Mcf) was used to estimate the emissions from the CNG vehicles. The reference case emissions were calculated from the gasoline equivalent of the natural gas consumed using the emission rate for motor gasoline (19.564 lbs/gal). The latter emissions represent the emissions that would have occurred if the vehicles had been operated with gasoline. The following reduction estimates reflect both Central Hudson vehicles (direct CO2 reductions) and non-Central Hudson vehicles (indirect CO2 reductions) that refuel at the Central Hudson refueling station. 2000 CO2 Reduction Calculation: Natural Gas: 1,051.96 Mcf (2000)x 120.593 lbs CO2/Mcf = 126,859.0123 lbs CO2 126,859.0123 lbs CO2/2,000 lbs per short ton = 63.430 tons CO2 Gasoline: 8,852.25 gal displaced(2000)x 19.564 lbs CO2/gal = 173,185.419 lbs CO2 173,185.419 lbs CO2/2,000 lbs per short ton = 86.593 tons CO2 Reduction: 86.593 tons CO2 - 63.430 tons CO2 = 23.163 CO2 ton reduction

Cinergy Corp.

Fleet Alternative Fuels		Direct: 108.64
Project Description	131 vehicles	
	The Cinergy Corp. operates a certain number of its vehicles using the alternative fuels propane and natural gas. The company has one propane filling station and currently has three natural gas filling stations (two open to the public). The natural gas vehicles are dual fuel vehicles - natural gas and gasoline. This is due to the fact that compressed natural gas is used and has a limited volume, which limits vehicle range. Propane is used in passenger vehicles, light trucks, and heavy trucks. Compressed natural gas is used in passenger vehicles and light trucks. The company has an aggressive program to provide technical assistance and compressor equipment to other fleet operators, and has opened a commercial conversion facility for the general public. Emissions reported for this project are emissions for the entire vehicle fleet, based on motor gasoline, diesel, propane and natural gas consumption.	
Estimation Method	The following were the emission rates used, all from Instructions, Appendix B: 19.641 lb CO2/gal gasoline 12.669 lb CO2/gal propane 120.593 lb CO2/Mcf natural gas	

Conectiv Delmarva Generation

CNG Vehicles		Direct: 29.94
Project Description	29 vehicles	
	Vehicles run on compressed natural gas (CNG) instead of gasoline. Beginning 1995, external fleets will also operate on natural gas. However, reductions reported in Part III reflect Delmarva Power's vehicles only.	

| Estimation Method | For 2000: CO_2 (tpy) = # CNG vehicles x (12,734 miles/yr)/(24 miles/gallon) x [19.6 lb CO_2/gal gasoline - (120.6 lb CO_2/mscf NG x 0.127 mscf NG/gal gasoline)]/2000 lbs/st) |

Entergy Services, Inc.

| Natural Gas Vehicle Program | 62 vehicles | Direct: 101.60 |

| Project Description | The natural gas vehicles program began in Baton Rouge, La in 1981 and in New Orleans, LA in 1993. |

Estimation Method	The net CO_2 reductions from using natural gas instead of gasoline to fuel vehicles was calculated as follows:
	CO_2 Emissions decreased (tons) = gasoline displaced (gallons) * 19.564 lbCO_2/gal * 1/2000 tons/lbs
	CO_2 emission increase from use of natural gas(tons) = natural gas used (Mscf) * 120.593 lbCO_2/Mscf * 1/2000 tons/lb
	Net CO_2 reductions = CO_2 Emissions decrease - CO_2 Emissions increase

Niagara Mohawk Power Corporation

| Alternative Fuel Vehicles | 30 vehicles | Direct: 22.04 |

| Project Description | NMPC has been involved in operating and testing alternative fuel vehicles (AFVs) for almost 30 years. The Company also currently has a number of "Clean Air" natural gas-fueled buses in operation as part of a cooperative program with the Syracuse, New York Centro transit system. |

| Estimation Method | CO_2 emission reductions are based on the difference in CO_2 emissions between gasoline-fueled vehicles and CNG or electric vehicles. Only direct emission reductions are reported. Emissions estimates are based on a CO_2 emission factor for each fuel. For motor gasoline, an emission factor of 19.641 lbs/gallon was used. For diesel fuel, an emission factor of 22.384 lbs/gallon was used. For CNG vehicles, a factor of 120.593 lbs/Mcf was used. These factors are based upon Form EIA-1605, Voluntary Reporting of Greenhouse Gases, Instructions, Appendix B. Fuel and Energy Source Codes and Emission Coefficients: EIA, 1996. For electric vehicles, NYPPs marginal emissions rate of 1.44 lbs/kWh for the years 1991-1995, rate of 1.48 lbs/kWh for 1996, and 1.46 lbs/kWh for 1997 and 1998 were used. These marginal rates were determined based on production simulation modelling (PROMOD IV). |

NiSource/NIPSCO

| Natural Gas Vehicles | 600 vehicles | Direct: 646.82 |

Project Description	PROJECT DESCRIPTION: NIPSCO is committed to significantly increasing the percentage of NGVs in our fleet over the next several years through the following actions:
	1) Purchasing factory-direct dedicated NGVs as available
	2) Converting forklifts and light duty vehicles and trucks to compressed natural gas (CNG)
	3) Utilizing liquified natural gas (LNG) in our heavy duty trucks.

	In addition to utilizing natural gas in our own fleet, NIPSCO will increase the number of NGVs operating throughout our region by providing a highly reliable fueling infrastructure, and by developing strategic alliances with educational, governmental, and social organizations. NGV training and consulting services will be provided to meet the mandates of the Clean Air Act Amendments of 1990.

NIPSCO has been a leader in NGVs since 1981. The NGV market is expected to increase (in accordance with mandates contained in the Energy Policy Act of 1992). Market demands and preferences will then drive the further proliferation of NGVs through the end of this century. |
| **Estimation Method** | NIPSCO used the following data below in its calculations.

Assumptions: for 1994-1998
CNG fuel usage rate = 46,886,000 cf/569 vehicles = 82,400 cf/vehicle
HHV gasoline 125,000 BTU/gal
HHV Natural Gas = 1030 BTU/cu.ft
Conversion Factor from NG to gasoline = (125,000 BTU/gal gas)(1 cu.ft. NG/1030 BTU)(1 BTU NG/0.94 BTU gas)
Conversion Factor from NG to gasoline = 129.1 cu.ft. NG/gallon of gasoline

Emission factors:
NG = 0.1206 lbs CO2/cu.ft.
Gasoline = 19.64 lbs CO2/gallon

Calculations:
Calc.1   (Number of CNG vehicles) x 82,400 cf/vehicle = cu.ft. NG
Calc.2   (cu.ft. of CNG) / 129.1 cu. ft. NG/gallon of gasoline = equiv. gallons of gasoline
Calc.3   (cu.ft. NG) x 0.1206 lbs CO2/cu.ft./2000 = tons CO2 from NG
Calc.4   (gallons of gasoline) x 19.641 lbs CO2/cu.ft./2000 = tons CO2 from gasoline
Calc.5   Difference between NG CO2 and gasoline CO2

For 1999 NiSource implemented an automated fuel tracking system and was able to more accurately report the amount of GGE (Gallons of Gasoline Equivalent) used throughout our service territory. For 1999 onwards, assume NG to gasoline conversion = 121 cu.ft. NG/gallon of gasoline.

Year	GGE	CO2 Emissions (tons)	CO2 Reductions (tons)
2000	286,696	2,091	713

1 GGE (gallon of Gasoline Equivalent) x 121 = cu ft NG

NG = 120.593 lbs CO2/1000 cu ft

1 GGE CO2 emission rate = 1 GGE*121*120.593 lbs CO2/1000 cu ft = 14.59 lbs CO2/GGE
1 Gallon of Gasoline CO2 emission rate = 19.564 lbs CO2/gallon gasoline |

	Savings = 19.564 - 14.59 = 4.974 lbs CO2/GGE	
PECO Energy Company		
Operation of CNG Vehicles	21 dedicated vehicles and 43 bifuel vehicles	Direct: 33.77 Indirect: 11.40
Project Description	PECO Energy's natural gas vehicle fleet used 12.768 gasoline gallon equivalents during year 2000. PECO has moved away from CNG use in passenger vehicles and is utilizing larger applications (e.g. pick up trucks). In 2000, PECO's CNG fleet was comprised of the following: Cars (2 dedicated CNG) Pickup trucks (1 dedicated CNG and 20 bifuel) Vans (16 dedicated CNG and 19 bifuel) Sport Utility (0 dedicated CNG and 4 bifuel) Floor Sweeper (1 dedicated CNG) Trailer (1 dedicated CNG) 21 dedicated CNG vehicles. 43 bifuel vehicles.	
Estimation Method	The total fleet mileage of PECO's dedicated and bifuel vehicles in 2000 was over 536,000 miles. Dedicated CNG vehicles traveled over 130,000 miles while bifueled vehicles traveled over 407,000 miles. PECO does not monitor the mileage use different fuels in the bifuel vehicles only total mileage. Therefore, for these purposes only the dedicated mileage was used. This is undoubtedly a conservative approach, as the PECO bi fuel vehicle mileage is not accounted for. The quantity of natural gas burned in 2000 (1,404.497 MSCF) is assumed to displace 12.768 gallons of gasoline. This was computed using the assumption that 0.11 MSCF of CNG equals one gallon of gasoline. Indirect and direct emissions and reductions were calculated from: Annual Emissions = Annual Mileage*FM + Annual Fuel Use*Ff Where: FM = emissions factor per mile driven (from EIA instructions) Ff = emissions factor per unit of fuel used (from EIA instructions)	
PG&E Corporation		
Natural Gas Vehicles	501 vehicles	Direct: 5091.12
Project Description	Pacific Gas and Electric Company Clean Air Vehicle Program: In 1990 Pacific Gas and Electric Company received California Public Utility Commission approval to spend up to $50 million by December 31, 1994 to support the development and introduction of electric and natural gas vehicles. By the end of 1993, Pacific Gas and Electric Company was operating 698 natural gas vehicles and 30 natural gas refueling stations. Encouragement took many forms: demonstrating vehicle and station performance, providing natural gas refueling station designs, providing partial funding for vehicle purchases, opening Company stations for public use, etc. After 1994, there was a decreased emphasis on customer	

	financial support. But Pacific Gas and Electric Company has continued to promote, facilitate and encourage electric and natural gas vehicle use by its customers. Pacific Gas and Electric Company continues to claim credit for not only its own fuel displacement, but also for displacements that it has encouraged its customers to undertake.	
Estimation Method	Natural gas therms used by natural gas vehicles is estimated from meter records of natural gas delivered by Pacific Gas and Electric Company to its own natural gas vehicle refueling stations, and of the natural gas supplied to customer owned natural gas refueling stations within its service territory. Pacific Gas and Electric Company takes credit for natural gas savings by customers within the Company's northern and central California service territory because Pacific Gas and Electric Company ratepayers funded a comprehensive program to promote natural gas use in vehicles, which program included both financial and technical support for numerous customer stations. Using the following factors, the Company calculates CO_2 emissions and emissions avoided through displaced gasoline: 103,001 mmBtu per million therms 1.1 therms per equivalent gallon of gasoline 117.08 lbs. CO_2 per mmBtu natural gas 19.564 lbs. CO_2 per gallon of gasoline In 1999 a total of 7.065 million therms of natural gas were used to displace gasoline. 62,827 tons CO_2 gasoline - 42,599 tons CO_2 natural gas = 20,228 tons CO_2 avoided. A similar methodology was applied to year 2000 data.	

Portland General Electric Co.

Natural Gas Fleet Vehicles	312,000 vehicle miles traveled	Direct: 54.59
Project Description	These are fleet vehicles voluntarily converted to natural gas. They operate in PGE's service area and commute to generation facilities. This area is the northern Willamette Valley and Columbia River gorge.	
Estimation Method	We know that 2 vehicles were converted in 1993 and 4 additional vehicles were converted in 1994. Fifteen more natural gas vehicles were delivered in mid-year (June) 1997. In 1998 eight 1/2 ton pickups were converted allow natural gas as a fuel in mid-year 1998. In 1999, another ten 1/2 ton pickups were converted to allow natural gas as a fuel in mid year 1999. No new vehicles were converted in year 2000, but all converted vehicles were operating for the full year. We assume the fleet vehicles travel 8000 mi/year each, that the gasoline mileage is 20 mi/gal, and that each gasoline vehicle emits 7838 pounds of CO_2 per year and each NG vehicle emits 4752 pounds per year. Fuel use for the NG vehicle was estimated using a conversion of 118 pounds of CO_2 per MBTU of energy.	

Public Service Company of New Mexico

CNG Vehicles	N/A	N/A
Project Description	PNM has been increasing the use of CNG vehicles in its fleet, particularly for its cars and small trucks and vans. In the twelve-month period ending 6/30/97, PNM vehicles logged nearly 24 million miles. Of this amount, CNG-capable vehicles logged 4,082,778 miles, of which approximately 90% of these miles were fueled by CNG (the balance were fueled by unleaded gasoline as the vehicles are dual fueled). Since CNG is a lower carbon fuel than is gasoline, approximately 40 pounds of CO_2 are saved for each MMBtu of gasoline displaced. This is based on emission	

factors of 157.041 lbs CO2/MMBtu for motor gasoline and 117.080 lbs CO2/MMBtu for natural gas.

1997: In the period 7/1/97 through 5/31/98, PNM fleet vehicles logged approximately 13 million miles. Of this amount, CNG-capable vehicles logged 1,964,763 miles of which approximately 75% of these miles were fueled by CNG (the balance were fueled by unleaded gasoline as the vehicles are dual fueled). Since CNG is a lower carbon fuel than gasoline, approximately 41.4 pounds of CO2 are not released for each MMBtu of gasoline displaced. This is based on emission factors of 156.662 lb CO2/MMBtu for motor gasoline and 115.258 lb CO2/MMBtu for natural gas (1998 Instructions for Form EIA-1605, Appendix B).

1998: Since the last reporting period, PNM fleet vehicles logged approximately 11.9 million miles. Of this amount, CNG-capable vehicles logged 1,355,833 miles. Since CNG is a lower carbon fuel than gasoline, approximately 41.167 pounds of CO2 are not released for each MMBTU of gasoline displaced. This is based on emission factors of 156.425 lb CO2/MMBTU for motor gasoline and 115.258 lb CO2/MMBTU for natural gas (1999 Instructions for Form EIA -1605, Appendix B).

1999 & 2000:   Data not available.

Estimation Method	

1964763   miles in CNG vehicles
X 75%   of those miles CNG-fueled
= 1,473,572 CNG miles
/12   miles per gallon equivalent
= 122,798 gallons equivalent of CNG used
/ 8.08   gallons per MMBtu
= 15,198 MMBtu of CNG used
X 41.4   lb CO2 saved per MMBtu (see note below)
= 629,197 lb of CO2 not emitted
=314.6   tons CO2 not emitted

NOTE:

MG = 156.662 lb CO2/MMBtu
CH4= 115.258 lb CO2/MMBtu
MG-CH4 = 41.4 lb CO2/MMBtu

1998
The CO2 savings are estimated as follows:
1,355,833 miles in CNG vehicles
/12   miles per gallon equivalent
=112,986 gallons equivalent of CNG used
/8.08   gallons per MMBtu
= 13,983   MMBTU of CNG used
X 41.167 lb CO2 saved per MMBTU (see note below)
= 575,638 lb CO2 not emitted
= 287.8   tons of CO2 not emitted

NOTE: MG = 156.425 lb CO2/MMBTU
CH4 = 115.258 lb CO2/MMBTU
MG-CH4 = 41.167 lb CO2/MMBTU

The first set of available calculations is for 1996.

1999 & 2000:   Data not available.

Tennessee Valley Authority

Alternate Fuel Vehicles	N/A	N/A

Project Description

In 1994, TVA had 31 alternate fuel vehicles operating in its transportation fleet. These included 23 sedans fueled by M-85 (a blend of 85% methanol and 15% gasoline), 2 compressed natural gas vans, 5 electric pickup trucks, and one electric van.

In question 4, the alternate fuel type listed as "ZZ" is the M-85.

Project results for 1995, 1996, 1997, 1998, 1999 and 2000 are not reported as data were not available.

Estimation Method

The direct emissions shown in Part 3 are the emissions used to compute the reported emissions reductions. These are the total emissions from the TVA transportation fleet. The actual CO2 emissions were determined from the fuel consumed and the fuel emissions factor from Appendix B. See the previous project, Transportation Fleet Fuel Efficiency Improvements.

The CO2 reductions as a result of alternate fuel vehicle (AFV) operation is the net difference between the modified reference case CO2 emissions and the actual emissions from the AFVs. The modified reference case emissions are the emissions that would have occurred had the miles driven by the AFVs been driven by the conventional fleet. The modified reference case emissions were determined from the actual AFV miles traveled, the average miles per gallon for the comparable conventional vehicles, the heating value of gasoline (125,100 BTU/Gal), and the gasoline emissions factor from Appendix B (157 lb CO2/MM BTU). It was assumed that the electric and CNG vehicles displaced emissions from the conventional 4X2 pickup fleet and the M-85 vehicles displaced emissions from the conventional sedan fleet.

The actual emissions for the CNG and M-85 AFVs were determined from the fuel usage, the heating value of the fuel, and the fuel emissions factor. The heating value for CNG is 1000 BTU/Ft3 and for M-85 is 73,590 BTU/Gal. The emissions factor for CNG is 120 lbs CO2/MM BTU and 146 lbs CO2/MM BTU for M-85.

To determine the actual emissions for the electric vehicles it was assumed that the energy used to charge the vehicles was generated by the TVA coal fired system. The emissions associated with the charging were determined from the KWH used, the average coal fired system heat rate, and the coal emissions factor from Appendix B.

The following table summarizes the operation of the AFVs and the resulting effect on CO2 emissions for 1994. In this table, negative changes, i.e. reductions, are shown in parentheses.

Alt. Fuel	Change In Miles Driven	Alt. Fuel Used	Conv. Vehicle MPG	Change In Gasoline Gallons	Conv. Vehicle CO2 Tons	Heat Rate BTU/ KWH	Fossil Fuel CO2 Tons	Change in CO2 Emission Tons
M-85	14258	544 Gal	29.8	(478)	(4.7)	--	2.9	(1.8)
CNG	1301	25000 CF	15.5	(84)	(0.8)	--	1.5	0.7
Elec.	4201	1360 KWH	21.2	(198)	(1.9)	10047	1.4	(0.5)
TOTAL 19760				(760)	(7.5)		5.8	(1.6)

TXU

Alternative Fuel Vehicle Program	221 vehicles	Direct: 212.28
		Indirect: 151.50

Project Description

TXU operates a fleet of alternatively fueled vehicles (chiefly compressed natural gas). This is the fifth year that the Company has included the carbon dioxide emissions reductions from these vehicles in the Climate Challenge Program.

Estimation Method

Estimates of the reduction of carbon dioxide from operating alternative fueled vehicles were based on the assumption that equivalent miles would have been driven by gasoline powered vehicles. First, the equivalent tons of carbon dioxide from gasoline vehicles were calculated then this quantity was subtracted from the equivalent tons of carbon dioxide generated from alternative fueled vehicles driving the same number of miles. Emission factors for carbon dioxide per fuel type were taken from Tables 4.2 and 4.3, page 4.19 of the Sector-Specific Issues and Reporting Methodologies, Volume II, part 4- Transportation Sector, October 1994. The DOT CAFE Standard of 27.5 mpg divided by 1.15 was used as the miles per gallon of gasoline and 20 mpg divided by 1.15 for propane was estimated.

The emission factors used for this project are listed:

	direct	indirect	Total
gasoline	8.900	2.100	11,000 g/gal
Propane	5,747	483	6,230 g/gal
methane	60.5	3.9	64.4 g/ft3

Western Resources, Inc.

Conversion of Company Fleet Vehicles to Alternative Fuels	5 vehicles	Direct: 4.2

Project Description

Conversion of Company Fleet Vehicles to Alternative Fuel Vehicles - Western Resources has converted company fleet vehicles to compressed natural gas (CNG) or dual fuel (CNG and gas/diesel) vehicles. These alternative fuel vehicles emit approximately 1/2 of the equivalent CO2 emissions as conventional vehicles. Western Resources currently has 5 alternative fuel vehicles (AFV).

Estimation Method

(1)   Western Resources has converted the following fleet vehicles to alternative fuel vehicles:
1991 CNG Vehicles - 0   1991 Dual Fuel Vehicles - 3
1992 CNG Vehicles - 6   1992 Dual Fuel Vehicles - 20

1993 CNG Vehicles - 6   1993 Dual Fuel Vehicles - 15
1994 CNG Vehicles - 9   1994 Dual Fuel Vehicles - 16
1995 CNG Vehicles - 2   1995 Dual Fuel Vehicles - 22
1996 CNG Vehicles - 2   1996 Dual Fuel Vehicles - 22
1997 CNG Vehicles - 2   1997 Dual Fuel Vehicles - 22 (Jan. to Nov.1997)
11/97 CNG Vehicles - 1   1997 Dual Fuel Vehicles - 4 (December 1997)
1998 CNG Vehicles - 1   1998 Dual Fuel Vehicles - 4
1999 CNG Vehicles - 1   1999 Dual Fuel Vehicles - 4
2000 CNG Vehicles - 1   2000 Dual Fuel Vehicles - 4

In November 1997, 19 AFVs were transferred to OneOak, with the remaining 5 APVs being retained by Western Resources (1 CNG and 4 Duel Fuel).

(2)   Based on information available from Argonne National Laboratories studies, the overall equivalent CO_2 emissions reduction of a CNG vehicle compared to a conventional vehicle is approximately 1.05 metric tons annually. This includes the net effect of an equivalent reduction in N_2O emissions and an equivalent increase in CH_4 emissions. This emissions data was also summarized in the 1605(b) transportation guidelines. It is assumed company vehicles are used for 10,000 vehicle miles traveled (VMT) per year on average.

(3)   Assuming dual fuel vehicles are operated on CNG 75% of the time and therefore, reduce equivalent emissions by 75% of a dedicated CNG vehicle, the equivalent CO_2 emissions avoided are estimated as:

1991 Equiv. CO_2 Emissions Avoided = 2 metric tons
1992 Equiv. CO_2 Emissions Avoided = 22 metric tons
1993 Equiv. CO_2 Emissions Avoided = 18.1 metric tons
1994 Equiv. CO_2 Emissions Avoided = 22.1 metric tons
1995 Equiv. CO_2 Emissions Avoided = 19.4 metric tons
1996 Equiv. CO_2 Emissions Avoided = 19.4 metric tons
1997 Equiv. CO_2 Emissions Avoided = 18.2 metric tons
1998 Equiv. CO_2 Emissions Avoided = 4.2 metric tons
1999 Equiv. CO_2 Emissions Avoided = 4.2 metric tons
2000 Equiv. CO_2 Emissions Avoided = 4.2 metric tons

(4)   Future CO_2 emissions avoided were based on a continuation of the 4.2 metric tons in the future for the two years covering 2001-2002.

(5)   No indirect emissions impacts were estimated.

Wisconsin Electric Power Co.

Vehicle conversion to dual fuel capability	874 vehicles	Direct: 2,126.44
		Indirect: 1,941.18

Project Description	VEHICLE CONVERSIONS TO DUAL FUEL CAPABILITY
	Conversion of gasoline-fueled vehicles to dual fuel capability (gasoline and Compressed Natural Gas or CNG) reduces CO_2 emissions while the vehicle is using CNG. WE also has a Vehicle CNG Program in which they provide technical assistance to customers wishing to utilize CNG vehicles. Assistance includes:

	An assessment of how clean fuel legislation and requirements affect the customers business: identification of available technology: determination of the suitability of the customers fleet for conversion; calculation of the cost of the conversion; determination of operating cost savings; determination of fueling station requirements; calculation of payback; and acquisition of bids from conversion equipment vendors. WE will assist in facilitating cooperation between groups who may wish to share the cost of refueling equipment. WE also has a custom spreadsheet to evaluate rebate incentives for larger fleets. In addition, WE provides incentives to encourage conversion of customers vehicles and WE employees personal vehicles to CNG. These incentives include a rebate of $500 or $0.50 per annual therm, whichever is greater, for each vehicle converted up to two vehicles; available financing (at 0% interest for WE employees); and fueling availability at WE fueling stations. System CO2 emission reductions due to CNG vehicle conversions in baseline years (in tons): 1987 - 94 1988 - 89 1989 - 55 1990 - 49

Estimation Method	Data Source: Form 1605(b) instruction manual
	
	Calculations:
	CO2 Emission Reductions =
	   CO2 (gasoline saved) - CO2 (natural gas used) =
	   (gal. gasoline * emission factor) - (mscf * emission factor)
	
	Direct reductions are related to conversion of company vehicles. Indirect reductions are related to conversion of customer vehicles.
	
	NOTE: 1998 and revised 1997 values reflect the unavailability of CNG conversion kits for WE fleet vehicles plus an error in calculation that had resulted in understating total 1997 CO2 emission reductions by about 700 st.
	
	Emission rates were revised for 1995 through 1998 in 10/99.

* Project Size refers to size in 2000 unless otherwise noted.

^ Project Description and Estimation Method are quoted directly from the Reporters' 2000 EIA-1605 reports.

As stated in Chapter 1, there are more than one million natural gas cars, trucks, and buses operating worldwide, with nearly 4,000 refueling stations to support the vehicles. The vast majority of these vehicles are located in Argentina, Italy, the United States, Brazil, Russia, Venezuela, and Canada. Argentina leads the world with more than 400,000 NGVs followed by Italy with over 300,000, the United States with approximately 104,000 and Brazil with 60,000. Although the United States ranks third in terms of numbers of NGVs, it ranks first in the world in total number of refueling stations with over 1,200 nationwide. NGV technology is not new to the world. Italy has been using natural gas as a vehicle fuel since the 1920s. In the United States, NGVs have been in use since the 1960s and NGVs played an important role in the former Soviet Union's vehicle fleets. Moreover, countries such as Canada and Venezuela have national programs that provide assistance for vehicle conversion and refueling stations.

There are many U.S. companies that are heavily involved in NGV development internationally. The list includes Deere Power Systems Group, Cummings Engines, Natural Gas Vehicle Company, Dyntech Industries Inc., Thomas Built Buses, NGV EcoTrans, Pressed Steel Tank Company, Blue Energy Inc., and Hurricane Compressors.

Figure A1. Worldwide Distribution of NGVs

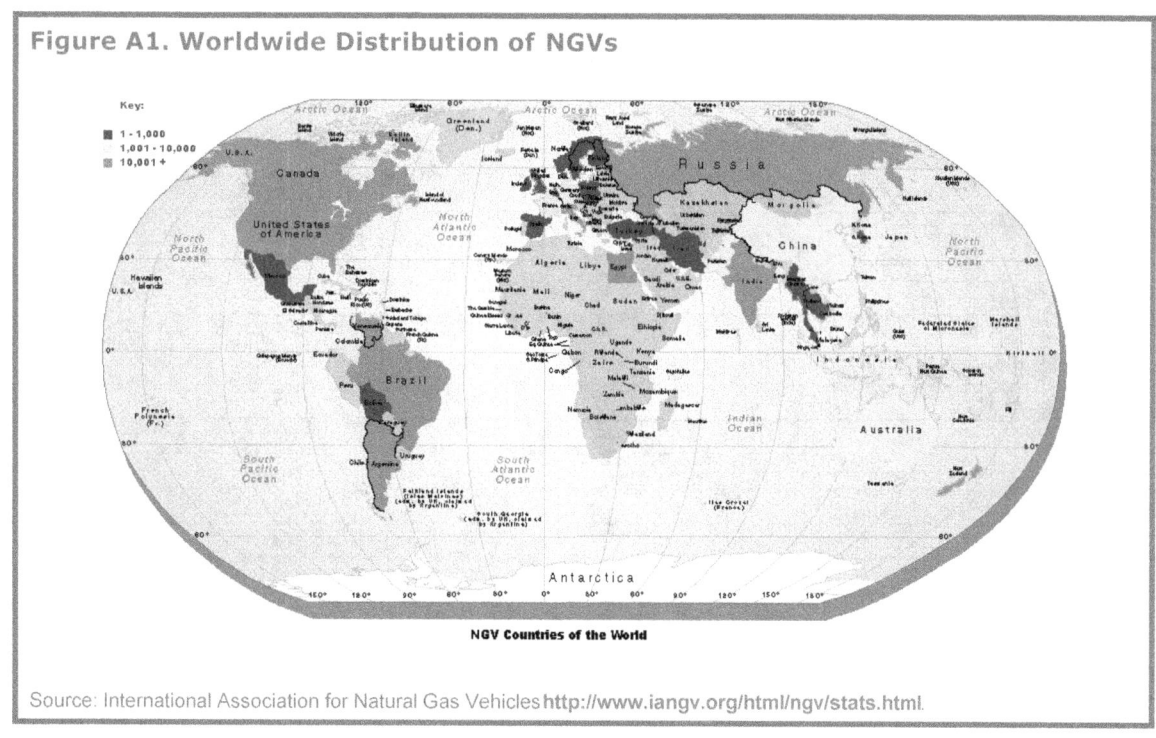

NGV Countries of the World

Source: International Association for Natural Gas Vehicles http://www.iangv.org/html/ngv/stats.html.

Table A2. International NGV Statistics for Vehicles Counted as of August 2000		
Country	Vehicles Converted	Refueling Stations
Argentina	462,186	830
Italy	320,000	320
United States	90,000	1,250
Brazil	60,000	55
Russia	30,000	208
Venezuela	27,542	151
Canada	20,505	222
Egypt	19,000	35
New Zealand	12,000	100
India	10,000	11
China	6,000	70
Japan	5,684	107
Germany	5,000	110
Bolivia	4,860	17
Colombia	4,500	22
Pakistan	4,000	30
Trinidad & Tobago	4,000	12
Malaysia	3,700	17
France	3,309	105
Indonesia	3,000	12
Chile	2,000	5
Sweden	1,500	22
Australia	1,000	35
Mexico	1,000	5
Bangladesh	1,000	5
Great Britain	835	18
Iran	800	2
Holland	574	27
Spain	300	6
Belgium	300	5
Switzerland	270	14
Burma	200	
Turkey	189	3
Austria	83	5
Thailand	82	1
Ireland	65	1
Finland	34	5
Czech	30	11
Nigeria	28	2
Luxembourg	25	5
South Korea	22	1
Poland	20	4
Norway	18	3

Denmark	5	1
Korea	4	1
Algeria		1
Totals	1,105,670	3,872

Source: International Association for Natural Gas Vehicles **http://www.iangv.org/html/ngv/stats.html**.

A4 U.S. Initiative on Joint Implementation (USIJI) Project Criteria

Criteria from the Final USIJI groundrules as published in the Federal Register on June 1, 1994:

"Section V—Criteria

A.   To be included in the USIJI, the Evaluation Panel must find that a project submission:

(1)   Is acceptable to the government of the host country;

(2)   Involves specific measures to reduce or sequester greenhouse gas emissions initiated as the result of the U.S. Initiative on Joint Implementation, or in reasonable anticipation thereof;

(3)   Provides data and methodological information sufficient to establish a baseline of current and future greenhouse gas emissions:

(a)   In the absence of the specific measures referred to in A.(2)-- of this section; and

(b)   As the result of the specific measures referred to in A.(2) of this section;

(4)   Will reduce or sequester GHG emissions beyond those referred to in A.(3)(a) of this section, and if federally funded, is or will be undertaken with funds in excess of those available for such activities in fiscal year 1993;

(5)   Contains adequate provisions for tracking the GHG emissions reduced or sequestered resulting from the project, and on a periodic basis, for modifying such estimates and for comparing actual results with those originally projected;

(6)   Contains adequate provisions for external verification of the greenhouse gas emissions reduced or sequestered by the project;

(7)   Identifies any associated non-greenhouse gas environmental impacts/benefits;

(8)   Provides adequate assurance that greenhouse gas emissions reduced or sequestered over time will not be lost or reversed; and

Provides for annual reports to the Evaluation Panel on the emissions reduced or sequestered, and on the share of such emissions attributed to each of the participants, domestic and foreign, pursuant to the terms of voluntary agreements among project participants.

B.   In determining whether to include projects under the USIJI, the Evaluation Panel shall also consider:

>   (1)   The potential for the project to lead to changes in greenhouse gas emissions elsewhere;

>   (2)   The potential positive and negative effects of the project apart from its effect on greenhouse gas emissions reduced or sequestered;

Whether the U.S. participants are emitters of GHGs within the United States and, if so, whether they are taking measures to reduce or sequester such emissions; and

Whether efforts are underway within the host country to ratify or accede to the United Nations Framework Convention on Climate Change, to develop a national inventory and/or baseline of greenhouse gas emissions by sources and removals by sinks, and whether the host country is taking measures to reduce its emissions and enhance its sinks and reservoirs of greenhouse gases."

A5 Fuel and Energy Source Emission Coefficients[189]

Table A7	Fuel and Energy Source Emission Coefficients		
	Emission Coefficients		
Fuel	**Pounds CO_2 per unit volume or mass**		**Pounds CO_2 per million Btu**
Petroleum Products			
Aviation Gasoline	18.355	per gallon	152.717
	770.916	per barrel	
Distillate Fuel (No. 1, No. 2, No. 4 Fuel Oil and Diesel)	22.384	per gallon	161.386
	940.109	per barrel	
Jet Fuel	21.095	per gallon	156.258
	885.98	per barrel	
Kerosene	21.537	per gallon	159.535
	904.565	per barrel	
Liquified Petroleum Gases (LPG)	12.805	per gallon	139.039
	537.804	per barrel	
Motor Gasoline	19.564	per gallon	156.425
	822.944	per barrel	
Petroleum Coke	32.397	per gallon	225.130
	1356.461	per barrel	
	6768.667	per short ton	
Residual Fuel (No. 5 and No. 6 Fuel Oil)	26.033	per gallon	173.906
	1,093.384	per barrel	
Methane	116.376	per 1000 ft^3	115.258
Landfill Gas	a	per 1000 ft^3	115.258
Flare Gas	133.759	per 1000 ft^3	120.721
Natural Gas (Pipeline)	120.593	per 1000 ft^3	117.080
Propane	12.669	per gallon	139.178
	532.085	per barrel	

---

[189] Instructions for Form EIA-1505: Voluntary Reporting of Greenhouse Gases (for data through 2001). EIA Energy Information Administration, US Department of Energy. February 2002.

Electricity	Varies depending on fuel used to generate electricity[b]		
Electricity Generated from Landfill Gas	Varies depending on heat rate of the power generating facility		
Coal			
Anthracite	3,852.16	per short ton	227.400
Bituminous	4,931.30	per short ton	205.300
Subbituminous	3,715.90	per short ton	212.700
Lignite	2,791.60	per short ton	215.400
Renewable Sources			
Biomass	Varies depending on the composition of the biomass		
Geothermal Energy	0		0
Wind	0		0
Photovoltaic and Solar Thermal	0		0
Hydropower	0		0
Tires/Tire-Derived Fuel	6160	per short ton	189.538
Wood and Wood Waste [c,d]	3120	per short ton	195.000
Municipal Solid Waste [e]	1999	per short ton	199.854
Nuclear	0		0
Other	-		-

[a] For a landfill gas coefficient per thousand standard cubic foot, multiply the methane factor by the share of the landfill gas that is methane.

[b] For average electric power emission coefficients by state, see Appendix V (Previous Page).

[c] For as-fired dry wood

[d] Wood and wood waste contain "biogenic" carbon. Under international GHG accounting methods developed by the Intergovernmental Panel on Climate Change, biogenic carbon is considered to be part of the natural carbon balance and does not add to atmospheric concentrations of carbon dioxide.[190] Reporters may wish to use an emission factor of zero for wood, wood waste, and other biomass fuels in which the carbon is entirely biogenic.

---

[190] Intergovernmental Panel on Climate Change. *Greenhouse Gas Inventory Reference Manual: Revised 1996 IPCC Guidelines for National Greenhouse Gas Inventories*, Vol. 3, Pg. 6.28, (Paris France 1997).

A6　The Future of Alternative Fuel Vehicles

Although this manual focuses on the creation of GHG emission reduction projects using NGVs, many of the basic principles discussed for estimating baselines and additionality and documenting GHG emission reductions also apply to other transportation-related projects. While NGVs continue to be deployed at an increasing rate and offer substantial opportunities for reducing emissions in the transportation sector, there are also other technologies with the potential to meet medium- to long-term needs for transportation-related emission reductions.

The commercial deployment of NGVs continues to increase worldwide as issues including availability of re-fueling infrastructure, reduction of re-fueling time, and vehicle range, cost, and performance are resolved. As NGV use grows, governments and automobile manufacturers are also researching new types of AFV technologies and ways to improve existing technology. Other advanced technologies with the potential for use as GHG emission reduction projects include electric and hybrid electric vehicles, hydrogen fuel cell technologies, and gas-to-liquids, also known as "clean diesel." Though numerous other applications of AFVs are available, the above-listed technologies are the focus of this chapter due to their medium- to long-term potential to be used in addition to or as a replacement for NGVs. As more of these AFVs are manufactured, their cost is expected to drop due to economies of scale and resolution of many of the technological barriers.

A6.1　Electric Vehicles

Electric vehicles (EVs) operate much like traditionally fueled vehicles, except that they run on an electric motor instead of a combustion engine, and batteries instead of a fuel tank. Electricity is unique among the alternative fuels in that mechanical power is derived directly from it, whereas conventional fuels release stored chemical energy through combustion to provide motive power. Most often the electricity used to power EVs is provided by batteries. Researchers are also exploring the use of fuel cells to convert chemical energy to electricity, rather than relying on batteries for electricity storage (see Section 5.3 Hydrogen).

Since EVs can be recharged at home and/or at a fleet parking facility, they generally require no additional infrastructure, such as the building and/or modification of existing refueling stations. In addition, EVs are typically refueled during low-demand hours, so refueling is not limited by power supply. Assuming that vehicle manufacturers are able to bring down cost and increase vehicle range and that there are improvements in battery technology (see Section on Battery Types below), electric vehicles have the potential to become commercially deployed and serve as GHG emission reduction projects.

There are indications that certain applications of EVs may provide GHG emission reduction benefits of between 55 percent and 99.9 percent (CO_2 equivalent) depending on the energy source used for electricity generation.[191] Thus, provided that the source of electricity for refueling EVs is less carbon intensive than the full fuel-cycle CO_2 emissions from other transportation technologies, EVs have the potential to reduce the emissions and carbon intensity of the world's transportation sector.

A6.2   Hybrid-Electric Technology

Hybrid electric vehicles (HEVs) combine the internal combustion engine of a conventional vehicle with the battery and electric motor of an electric vehicle, resulting in twice the fuel economy of conventional vehicles.[192] This combination offers the extended range and rapid refueling that consumers expect from a conventional vehicle, and a significant portion of the energy and environmental benefits of an electric vehicle. The practical benefits of HEVs include improved fuel economy and lower emissions compared to conventional vehicles. The inherent flexibility of HEVs allows them to be used in a wide range of applications, from personal transportation to commercial hauling.

HEVs have the potential to significantly reduce GHGs due to factors including:

- Increased fuel efficiency (hybrids consume significantly less fuel than vehicles powered only by conventional fuels); and

- A reduction in dependency on fossil fuels because they can run on alternative fuels.

One of the most common forms of hybrid-electric technology is a heavy-duty application that combines an electric drivetrain with a diesel engine, which powers an alternator or generator to produce electrical power for heavy-duty application. This application is currently used by New York City Transit Authority and is being considered by other transit agencies in the United States. Decoupling the engine from the drivetrain allows it to be operated independently of vehicle speed. At a steady-state operating speed, a hybrid bus might be <u>less</u> fuel-efficient than the same bus using a conventional drivetrain. However, the real world driving conditions of the typical transit bus involves constant starts and stops. With a conventional drivetrain, the engine must be sized to provide sufficient power to accelerate the bus while operating all the needed accessories. A hybrid bus reduces the maximum power demand on the engine by recapturing braking energy and using it to help accelerate the bus from rest. This reduces the peak power requirement of the engine, allowing it to be smaller. By decoupling the engine from the drivetrain, further gains are possible by operating the engine only at its most efficient speeds and loads. Emissions are reduced, primarily as a function of reduced fuel consumption.

A current major concern with hybrid-electric buses is premature battery failure due to uneven charging. A partial remedy is an added maintenance step requiring charging of the batteries overnight or for an entire day to equalize their initial state of charge and operating voltage. At least once a month, this is recommended for hybrid buses in the field. Unless premature battery failure can be avoided, the cost of operating hybrid buses will be very high.

Hybrid buses are currently about 50 percent more expensive than conventional buses. Contributions to this increased expense include:

---

[191] Electric Vehicle Association of Canada, "Full Fuel Cycle Emission Reductions Through the Replacement of ICEVs with BEVs," July 10, 2000.
[192] DOE Clean Cities Website, "What is an HEV?" http://www.ott.doe.gov/hev/what.html.

- an electronic control system;

- a battery pack for energy storage;

- an electric drive motor; and

- recouping of R&D investments.

Hybrid busses are more expensive despite their smaller engines and simpler transmission systems. Also, because diesel engine emission standards are specified in terms of power output, any cost to comply with new engine emissions standards will apply equally to the engines used in hybrid and conventional buses.

A6.3   Battery Types

A large number of battery types are being tested for use in EVs. Some of the technologies under evaluation include lead-acid, nickel cadmium, nickel iron, nickel zinc, nickel metal hydride, sodium nickel chloride, zinc bromine, sodium sulfur, lithium, zinc air, and aluminum air.

In 1999, 1,277 battery light-duty vehicles were sold or leased in the United States. As of November 2000, an additional 476 battery light-duty vehicles were sold or leased in the United States. In both years, the Ford Ranger EV accounted for the majority of vehicles sold.

A6.4   Hydrogen

The lightest potential alternative fuel is hydrogen gas (H_2). Hydrogen is in a gaseous state at atmospheric pressure and ambient temperatures. Fuel hydrogen is not pure hydrogen gas, but rather contains small amounts of oxygen and other materials. H_2 is being explored for use in combustion engines and fuel-cell electric vehicles, although it presents greater transportation and storage hurdles than exist for the liquid fuels.[193] Storage systems being developed include systems designed for compressed hydrogen, liquid hydrogen, or a chemical bonding process between hydrogen and a storage material (for example, metal hydrides). Hydrogen is typically transported in canisters and tanker trucks. While no hydrogen-based distribution and refueling system is in place for the transportation sector, the ability to create the fuel from a variety of sources and its clean-burning properties make it a desirable alternative to conventional fuels.[194]

Two methods are generally used to produce hydrogen: (1) electrolysis and (2) synthesis gas production from steam reforming or partial oxidation. Electrolysis uses electrical energy to split water molecules into hydrogen and oxygen. The electrical energy can come from any electricity production source including renewable fuels. DOE has concluded that electrolysis is unlikely to become the predominant method for large quantities of hydrogen. The predominant method for producing synthesis gas today is steam-reforming of natural gas, although other hydrocarbons can be used as feedstocks. For example, biomass and coal can be gasified and used in a steam-reforming process to create hydrogen.

---

[193] Please note that researchers are investigating on-board reforming of liquid hydrocarbon or methanol for producing hydrogen for fuel cell-driven vehicles. This would avoid hydrogen storage problems.
[194] While pipeline transportation is generally the most economic means of transporting gaseous fuels, a pipeline system for hydrogen is currently not in place.

A6.5   Hydrogen Fuel Cells

Hydrogen fuel cells can be used as power generating systems for electric vehicles. They differ from battery-driven vehicles in that they store fuel, not energy. A hydrogen fuel cell works by converting the chemical energy of hydrogen and combining it with oxygen to produce electricity, heat, and water. The hydrogen is stored in tanks on board the vehicle, either as a liquid or as a gas. Fuel cell vehicles are still in the developmental stage, but with advances in technology, they may become viable.

A6.6   Clean Diesel

Clean diesel typically means diesel-fuel that is ultra-low in sulfur and nitrogen. Over the last few years, clean diesel has received much attention because it allows new power-train/fuel systems, such as fuel cells and ultra-clean diesel engines, to become reality. These new fuel systems will be necessary to meet the increasingly stringent clean air standards in the U.S. For example, the Environmental Protection Agency (EPA) recently introduced a rule requiring new pollution control devices to be effective on trucks and buses between 2007 and 2010 and mandating the sulfur content of highway diesel fuel to be reduced from its current level of 500 parts per million to 15 parts per million by 2006.[195]

New compression ignition (CI) engines are under development to meet the increasing dual challenges of greater fuel efficiency and reduced emissions of environmental pollutants. In particular, low-emission diesel engines are attractive because of their inherent 40% increase in fuel efficiency compared to gasoline engines. However, diesel engines are beginning to reach the limit of their performance envelope without substantial fuel improvements. The catalytic converters required to reduce oxides of nitrogen (NOX) emissions can not be used at present because the high sulfur levels (300 ppm) in the currently available fuels rapidly poison the catalyst of these anti-pollution devices. Ultra-clean diesel fuels could offer a way for these new vehicles to meet the more stringent emission standards without compromising safety, performance, or affordability.

A6.7   Gas-to-Liquids (GTL) Synthetic Fuel

The most promising method for producing clean diesel is the gas-to-liquids (GTL) synthetic fuel produced through the Fischer-Tropsch (FT) process. Synthetic fuel in the diesel range is made from natural gas using any of several FT processes. Unlike liquefied natural gas processing where natural gas is cooled to form a liquid, GTL technologies chemically change the natural gas molecules, breaking them apart, and re-combining them with oxygen to form a mixture called synthesis gas. In turn, synthesis gas can be chemically converted into different types of hydrocarbon products like clean-burning transportation fuels (clean diesel) or a variety of high-value chemicals. This conversion into liquid hydrocarbons (FT liquids) takes place on a Fischer-Tropsch catalyst. The synthetic fuel created through this process contains no detectable sulfur, aromatics, olefins, or alcohols. By eliminating these undesirable species, diesels can be made to operate as cleanly as gasoline or CNG engines, without penalizing efficiency.

One of the potential uses for GTL technology and clean diesel is as a replacement fuel for conventional diesel or as a blending agent with conventional fuels to help meet more stringent environmental regulations. Diesel with an ultra-low sulfur content is needed for emission control devices to reduce emissions of NOx and particulates effectively. Sulfur

---

[195] "EPA Dramatically Reduces Pollution From Heavy-Duty Trucks and Buses; Cuts Sulfur Levels in Diesel Fuel" Press Release, U.S. Environmental Protection Agency (EPA). Washington, DC. December 21, 2000. http://www.epa.gov.

is a major impediment to implementation of the emission control technology needed in diesel engines and can even cause increased particulate emissions when used with advanced catalytic particulate control devices designed to reduce emissions.[196]

Another use for clean diesel is as a fuel source for fuel cells. Both industry and government researchers have focused on conventional gasoline and diesel as fuel cell fuels because they can be delivered using the present fuel distribution system and have relatively high hydrogen carrying capacity. Like the diesel engine manufacturers, fuel cell fuel technology providers also prefer fuel with not detectable sulfur. As a result, clean diesel provides a useful fuel alternative that will help advance the development and utilization of fuel cells.

Compared to the obvious benefits on local air pollutants, the greenhouse gas emissions benefits of GTL technology are not as evident. The driving force behind GTL as a transportation fuel is that emissions of sulfur, NO_x, and particulate matter are significantly reduced compared to conventional transportation fuels. In terms of greenhouse gas emissions, there is no difference between conventional and new diesel—except in cases where the use of clean diesel allows for the use of a more fuel-efficient engine. However, there is a secondary effect to GTL that directly contributes to the reduction of greenhouse gas emissions; that is, GTL provides a use for natural gas that otherwise would have been flared. In this context, an argument could be made that a project utilizing GTL technology presents an alternative to natural gas flaring which results in greenhouse gas emission reductions.

There are approximately 12 GTL projects worldwide only two of which (both in South Africa) are currently operational. The remaining projects are considered potential and located in the United States, Venezuela, the United Kingdom, Nigeria, Norway, Qatar, Bangladesh, and Malaysia.[197] Despite the small number of projects, capital costs for a GTL project are becoming competitive with those associated with refining processes for conventional transportation fuel technologies. The U.S. Department of Energy estimates that the costs of the chemical conversion process could be reduced by 25 percent if a one-step process can be developed to separate oxygen from the air and combine it with natural gas to form synthesis gas. It would bring gas-to-liquid technology into the $18 to $20 per barrel range, which is competitive with crude oil.[198]

GTL also has unique economic advantages over other alternative fuel technologies on the distribution and end-use sides. First, GTL technologies yield products that can be used directly as fuels or feedstocks or they can be blended with crude oil products to help

---

[196] Wendy Clark, *et al*, "Overview of Diesel Emission Control—Sulfur Effects Program," SAE Paper 2000-01-1876 Presented at the CEC/SAE International Spring Meeting. Paris, France, June 19-22, 2000.

[197] Mark A. Agee, President and CEO, Syntroleum Corporation, Tulsa, OK "Fuels for the Future," Paper presentation at the *Energy Frontiers International Conference, Gas Conversion: Projects, Technologies, & Strategies*. San Francisco, CA October 20-22, 1999, http://www.syntroleum.com/pdfs/sf_1099.pdf.

[198] *DOE Fossil Energy Techline*, "University of Alaska-Fairbanks to Lead University, Industry Team, in Department of Energy Project to Develop 'Gas-to-Liquids' Technology," April 16, 1999, http://www.fe.doe.gov/techline/tl_akgastoliq.html.

comply with more stringent environmental requirements. Second, use of GTL fuels would not necessitate the rebuilding of vehicle fleets and distribution systems. GTL fuels could be delivered through existing infrastructures and existing vehicles would not necessarily need extensive modifications.[199] Other alternative fuels like CNG require new distribution systems, fueling stations, vehicle modifications, and cannot be blended with other crude oil products.

---

[199] Mark A. Agee, President and CEO, Syntroleum Corporation, Tulsa, OK, "Economic Gas To Liquids Technologies—A New Paradigm for the Energy Industry," Paper presentation at *Montreux Energy Roundtable VIII*. Montreux, Switzerland, May 12-14, 1997, http://www.syntroleum.com/pdfs/montreux597.pdf and "GTL vs. Low Oil Prices—The Insulating Factors." Paper Presentation at *Monetizing Stranded Gas Reserves '98 Conference*. San Francisco, CA, December 14-16, 1998, http://www.syntroleum.com/pdfs/sf_1298.pdf.

References

Agee, Mark A., "Economic Gas To Liquids Technologies – A New Paradigm for the Energy Industry." Paper presentation at *Montreux Energy Roundtable VIII*, Montreux, Switzerland, May 12-14, 1997, http://www.syntroleum.com/pdfs/montreux597.pdf.

Agee, Mark A., "Fuels for the Future." Paper presentation at *Energy Frontiers International Conference, Gas Conversion: Projects, Technologies, & Strategies*, San Francisco, CA October 20-22, 1999, http://www.syntroleum.com/pdfs/sf_1099.pdf.

Agee, Mark A., "GTL vs. Low Oil Prices – The Insulating Factors." Paper Presentation at *Monetizing Stranded Gas Reserves '98 Conference,* San Francisco, CA, December 14-16, 1998, http://www.syntroleum.com/pdfs/sf_1298.pdf.

American Automobile Manufacturers Association (AAMA), "World Motor Vehicle Data 1993," AAMA, (Washington, DC, 1993).

American Automobile Manufacturers Association (AAMA), "Motor Vehicle Facts and Figures 1996," AAMA, (Washington, DC, 1996).

Austin, T., R. Dulla and T. Carlson, "Alternative and Future Fuels and Energy Sources For Road Vehicles," Sierra Research Inc, 8 July 1999, http://www.tc.gc.ca/envaffairs/subgroups/vehicle_technology/study2/Final_r port/Final_Report.htm.

Bechtold, Rich, "Is Clean Diesel Fuel an Option for Chile?" Paper Presented at the International Seminar *Public Transport, Natural Gas, and the Environment: Challenges for 2000*, Santiago, Chile, May 22-24, 2000.

University of California at Irvine, Institute of Transportation Studies, "Choice Stated Preference Survey," Report UCT-ITS-WP-91-8, 1991.

California Air Resources Board, "California Exhaust Emissions Standards and Test Procedures for 2001 and Subsequent Model Passenger Cars, Light-Duty Trucks, and Medium-Duty Vehicles." Proposed Amendments, September 2001.

California Air Resource Board, *Low-Emission Vehicle Program*. September 28, 2001. http://www.arb.ca.gov/msprog/levprog/levprog.htm.

California Air Resources Board, "California's Zero Emission Vehicle Program Fact Sheet," December 6, 2001. http://www.arb.ca.gov/msprog/zevprog/factsheet/evfacts.pdf.

California Air Resources Board, "Notice of Public Hearing to Consider the Adoption of Amendments to the Low-Emission Vehicle Regulations," The California Low-Emission Vehicle Regulations, November 15, 2001.

California Energy Commission, "Global Climate Change and California." http://www.energy.ca.gov/global_climate_change/index.html.

Center for Biological Diversity v. Abraham, N.D. Cal., No. CV-00027. 2 January 2002.

Chomitz, Kenneth M., "Baselines for Greenhouse Gas Reductions: Problems, Precedents, Solutions," Draft Paper, World Bank (Washington D.C., July 1998).

Clark, Wendy, et al, "Overview of Diesel Emission Control—Sulfur Effects Program," SAE Paper 2000-01-1876. Presented at the *CEC/SAE International Spring Meeting*, Paris, France, June 19-22, 2000.

The Clean Air Act as Amended in 1990, http://www.epa.gov/oar/caa/contents.html.

The U.S. Climate Change Action Plan, (Washington, D.C., October, 1993), http://www.gcrio.org/USCCAP/toc.html.

Doherty, J. and Jette Findsen, "Case Study: CNG Taxis, The Republic of Clean Cities." Presentation at the session, *Developing International Greenhouse Gas Emission Reduction Projects Using Clean Cities Technologies*, San Diego, CA, May 10, 2000.

Cuenca, R.M., L.L.Gaines, and A.D. Vyas, "Evaluation of Electric Vehicle Production and Operating Costs," Argonne National Laboratory, (Chicago, November 1999).

EarthVision Environmental News, "New York Adopts New California Emission Standards," November 2002, http://www.climateark.org/articles/2000/4th/nyadnewc.htm.

Electric Vehicle Association of the Americas , "State Laws and Regulations Impacting Electric Vehicles," January 2002, http://www.evaa.org.

Energy Information Administration, "Instructions for Form EIA-1505: Voluntary Reporting of Greenhouse Gases (for Data through 2001)," February 2002, http://www.eia.doe.gov.

Energy Information Administration, "Updated State- and Regional-level Greenhouse Gas Emission Factors for Electricity," March 2002, http://www.eia.doe.gov/oiaf/1605/e-factor.html.

Energy Information Administration. Voluntary Reporting of Greenhouse Gases Program, http://www.eia.doe.gov/oiaf/1605/frntvrgg.html.

Federal Register, "Electric Vehicles," Volume 61, Number 51, pages 10627-10628.

Fulton, L, C. Liliu, M. Landwehr, and Lee Schipper, "Saving Oil and Reducing CO_2 Emissions in Transport: Options and Strategies," International Energy Agency, 2001, http://www.iea.org.

Hargrave, Tim, *et al*, "Options for Simplifying Baseline Setting for Joint Implementation and Clean Development Mechanism Projects," Center for Clean Air Policy (Washington, D.C., November 1998).

Intergovernmental Panel on Climate Change, "Greenhouse Gas Inventory Reference Manual: Revised 1996 IPCC Guidelines for National Greenhouse Gas Inventories," Vol. 3, 1997, http://www.unfccc.org.

International Association for Natural Gas Vehicles, "International Natural Gas Vehicle Statistics 2000 Online," http://www.iangv.org/html/ngv/stats.html.

International Energy Agency (IEA) and Organization for Economic Cooperation and Development (OECD), "Good Practice Greenhouse Abatement Policies: Transport

Sector." Papers prepared for the *Annex I Expert Group on the UNFCCC.* IEA and OECD, November 2000, http://www.iea.org.

Maryland Green Buildings Council, "2001 Green Buildings Council Report," November 2001, http://www.dgs.state.md.us/GreenBuildings/Documents/FullReport.pdf.

Massachusetts Low Emission Vehicle Program, Public Hearings on the Amendments to the State Implementation Plan for Ozone and Hearing and Findings under the Massachusetts Low Emission Vehicle Statute - 310 CMR 7.40, February 2002, http://www.state.ma.us/dep/bwp/daqc/daqcpubs.htm.

McCarthy, J. and S. Turner, "Natural Gas Vehicles and Greenhouse Gas Emissions." Presentation at the session, *Developing International Greenhouse Gas Emission Reduction Projects Using Clean Cities Technologies*, San Diego, CA, May 10, 2000.

National Oceanic and Atmospheric Administration, "Clear Skies & Global Climate Change Initiatives," February 14, 2002, http://www.whitehouse.gov/news/releases/2002/02/20020214-5.html.

Office of the Governor of New York, "Regulation to Reduce Harmful Vehicle Emissions, Alternative to Promote Clean Vehicle Technology, Improve Air Quality," January 4, 2002, http://www.state.ny.us/governor/press/year02/jan4_02.htm.

Pacific Northwest Pollution Prevention Center, "Alternative Fuels for Fleet Vehicles," http://www.pprc.org.

Pianin, Eric, "U.S. Aims to Pull Out of Warming Treaty: 'No Interest' in Implementing Kyoto Pact, Whitman Says," *Washington Post*, Wednesday, March 28, 2001.

Rosenzweig, R., M. Varilek, B. Feldman, R. Kuppalli and J. Jansen, "The Emerging International Greenhouse Gas Market," Pew Center on Global Climate Change, March 2002, http://www.pewclimate.org.

State of Maine Department of Environmental Protection, *Rule Chapter 127, New Motor Vehicle Emission Standard,* Basis Statement for Amendments of December 21, 2000.

Tompkins, M., et al., "Determinants of Alternative Fuel Vehicle Choice in the Continental United States," Transportation Research Record No. 1641, Transportation Research Board, (Washington, DC, 1998).

University of California at Irvine, Institute of Transportation Studies, "Choice Stated Preference Survey," Report UCT-ITS-WP-91-8, 1991.

United Nations Framework Convention on Climate Change (UNFCCC), http://www.unfccc.int/resource/conv/conv_002.html.

United Nations Framework Convention on Climate Change, *Activities Implemented Jointly*, http://www.unfccc.de/program/aij/aijproj.html.

United Nations Framework Convention on Climate Change, AIJ Uniform Reporting Format: Activities Implemented Jointly under the Pilot Phase, *The RABA/IKARUS Compressed Natural Gas Engine Project*, http://www.unfccc.int/program/aij/aijact/hunnld01.html.

U.S. Department of Energy, "Alternative Fuel Vehicle Fleet Buyer's Guide," http://www.fleets.doe.gov/cgi-in/fleet/main.cgi?17357,state_ins_rep,5,468050.

U.S. Department of Energy, "Federal Fleet AFV Program Status," June 2, 1998, http://www.ccities.doe.gov/pdfs/slezak.pdf.

U.S. Department of Energy, Office of Energy Efficiency and Renewable Energy, "Just the Basics: Electric Vehicles, Transportation for the 21st Century," (Washington, D.C. January 2002).

U.S. Environmental Protection Agency, *Legislative Initiatives*. http://yosemite.epa.gov/globalwarming/ghg.nsf/actions/Legislative Initiatives.

U.S. General Services Administration, Office of Government-wide Policy. http://policyworks.gov/org/main/mt/homepage/mtv/caaepact.htm.

U.S. Initiative on Joint Implementation, "Resource Document on Project & Proposal Development under the U.S. Initiative on Joint Implementation," Version 1.1, (Washington, DC, June 1997).

Wang, M., "Regulated Emissions and Energy Use in Transportation (GREET)," Argonne National Laboratory. http://www.transportation.anl.gov/ttrdc/greet.

Wisconsin Department of Natural Resources, *Clean Fuel Fleet Program*, http://www.dnr.state.wi.us/org/aw/air/reg/cff/cff.htm.

White House Office of the Press Secretary, "President Announces Clear Skies & Global Climate Change Initiatives," February 14, 2002, http://www.whitehouse.gov/news/releases/2002/02/20020214-5.html.

White House, "Global Climate Change Policy Book," (Washington, D.C., February 2002), http://www.whitehouse.gov/news/releases/2002/02/climatechange.html.

World Resources Institute, "Proceed With Caution: Growth in the Global Motor Vehicle Fleet," http://www.wri.org/trends/autos.html.

World Resources Institute and World Business Council for Sustainable Development, "The Greenhouse Gas Protocol: A Corporate Accounting and Reporting Standard," http://www.GHGprotocol.org.

World Resources Institute and World Business Council on Sustainable Development Greenhouse Gas Protocol Initiative, "Calculating CO_2 Emissions from Mobile Sources," http://www.GHGprotocol.org.